ADVANCED DRILLING SOLUTIONS

LESSONS FROM THE FSU

VOLUME 1

ADVANCED DRILLING SOLUTIONS

LESSONS FROM THE FSU

VOLUME 1

Yakov A. Gelfgat
Mikhail Y. Gelfgat
Yuri S. Lopatin

1421 South Sheridan Road
Tulsa, Oklahoma 74112-6600 USA

800.752.9764
+1.918.831.9421
sales@pennwell.com
www.pennwell-store.com
www.pennwell.com

Book design by Robin Remaley
Cover design by Amy Spehar
Managing Editor: Marla Patterson
Production Editor: Sue Rhodes Dodd

Library of Congress Cataloging-in-Publication Data

Gelfgat, Yakov A.
Advanced drilling solutions : lessons from the FSU /
by Yakov A. Gelfgat, Mikhail Y. Gelfgat, and Yuri S. Lopatin
p. cm.
Includes index
ISBN 0-87814-786-1
1. Oil well drilling--Russia (Federation) 2. Oil well drilling--Former Soviet Republics.
I. Gelfgat, Mikhail Y. II. Lopatin, Yuri S. III. Title.
TN871.2 G45 2003
622' .3382'0947--dc21 2002154490

Printed in the United States of America

1 2 3 4 5 07 06 05 04 03

CONTENTS

Volume 1

PREFACE

We are very happy to present this two-volume book to our respected readers. It is our hope that these volumes will be of interest to all professionals in the drilling industry. It is intended for those who work in the field, at the computer, as managers in the oil industry, university professors and students, and anyone who is interested in the history of the development of drilling technologies. We believe all readers will find something useful in these volumes, both in their current and future activities.

Three years ago we submitted a proposal to write this book for PennWell Publishing. That proposal was accepted, thanks to the unwavering encouragement and assistance of Dean Gaddy, a former drilling editor for the *Oil & Gas Journal.* We were also supported in this project commencing by the opinions of several drilling experts from different USA institutions—Dr. William Maurer (Maurer Engineering); Bill Gwilliam and Roy Long (DOE); Donald Dreesen (LANL) and Professor Stefan Miska from the University of Tulsa. We appreciate the PennWell team's endless patience and help in the preparation, editing, and formatting of the manuscript to the present form.

The objective of the book is to acquaint petroleum and drilling industry specialists with the well construction processes and new drilling technologies of the former U.S.S.R. and Russia. Following is the synopsis for these two volumes, which shows that the authors worked hard to create a book of interest to the oil & gas industry and to fulfill their commitment to PennWell.

Volume 1 of this series has three chapters, which cover historical trends and two major aspects of drilling technologies development in Russia—downhole motors and oil well drilling optimization.

Chapter 1 covers the goals of this volume and gives a detailed and comprehensive history of oil production in Russia. In the U.S.S.R., unlike in the United States and other countries primarily using rotary drilling, there are widely used downhole hydraulic and electrical motors, with oil and gas well footage drilled about 80%.

Chapter 2 discusses turbodrills, positive-displacement motors and electrodrill basics, design features, and operational results. The use of downhole motors brings substantial changes in borehole drilling and deepening technology as well as optimization methods development. The methods of directional and cluster drilling gained wide acceptance as far back as the World War II years, and at present are the dominant methods in Russia. The electrodrill application for drilling horizontal, branch, and other wells is one of the most important subjects presented in these volumes. This method has a unique advantage over hydraulic motor drilling; its performance is not dependent on the characteristics of the drilling fluid. Air and foams are widely used as circulation agents with electrodrills.

The so-called *Key technological wells* (KTW) drilling method, one of the most efficient methods of well construction optimization, is presented in Chapter 3. The chapter contains the mathematical model for the well deepening process as well as results of KTW technique application in development of several major oilfields in FSU. This chapter gives the rationale for three different drilling methods: rotary, hydraulic, and electrical downhole motors.

Volume 2 of this series consists of four other Chapters, which provide detailed descriptions and case studies of several technologies developed and widely used in Russia. There are directional drilling, deep and super-deep well construction, underbalanced drilling, rotary-turbine drills, underreamers, and retractable drill bits.

Chapter 4 covers all aspects of directional, cluster, horizontal and multi-lateral drilling technologies used in FSU and Russia. The details on the first horizontal and multi-lateral wells drilling in the world are given. Even though the FSU was the birthplace of horizontal drilling, the U.S.S.R. failed to bring the method to commercial application in the 1960s and 1970s, and now Russia is trying to catch up on modern technologies.

Chapter 5 presents the specific features of applications of downhole motors in super-deep drilling (6 to 7.5 km TVD), including rotary-turbine (RTB) drills used for large diameter vertical drilling. This chapter also contains the results of applications of under-reaming technology and the main results of ultra-deep drilling for scientific purposes. The world's deepest borehole, Kola SG-3 (12 km TVD), as well as Krivoy Rog SG-8 and others, were based on the utilization of downhole motors and aluminum drillpipe.

The underbalanced deep drilling technology is described in one section of chapter 5, as well as in chapter 6, which specifically discusses air, foam, and aerated mud drilling techniques, including booster pump technology features.

The development and usage, on land as well as offshore, of casing drilling (the original drilling with retractable bits without pulling out drillpipe), is the subject of consideration in Chapter 7.

The data presented in these two volumes is based on actual examples of carefully selected wells drilled in different time periods throughout the FSU/Russia producing areas—from West Ukraine to Far East and from Azerbaijan to North Europe and Siberia.

ACKNOWLEDGMENTS

Yakov A. Gelfgat wrote Chapters 3 and 4. He wrote Chapters 1, 2, and 5 with Mikhail Y. Gelfgat, who also wrote Chapter 7. Yuri S. Lopatin wrote Chapter 6. Boris Volkovoy translated the Russian text written by Yakov Gelfgat and Yuri Lopatin. The authors express their great appreciation for Volkovoy's efforts in finding solutions for difficult-to-translate technical text. Mikhail Gelfgat did the technical review of the English text.

The authors acknowledge the invaluable help of several prominent Russian drilling specialists in commenting, advising, and reviewing different book chapters and sections, namely:

Valeriy Petrovich Shumilov—section on turbodrills in Chapter 2, especially "multistage turbine theory development"

Dmitry Fedorovich Baldenko—section on positive displacement motors (PDM) in Chapter 2

Bairas Ibragimovich Abyzbaev—section on electrodrills in Chapter 2 and section on electrodrills application in directional drilling in Chapter 4

Bronislav Vasilievich Baidyuk—Chapter 3

Rudolf Stepanovich Alikin—under-reaming technology section in Chapter 5 and retractable bit design and application features sections in Chapter 7

Vladimir Solomonovich Basovich—section on the ultra-deep scientific drilling experience in Chapter 5

This manuscript could never have been delivered to the editor without significant support from the Aquatic Company staff in Moscow, especially that of Mrs. Elmira Minasovna Pogosyan, who typed most of the Russian text. We appreciate the help and efforts of our colleagues in Houston, Mr. Alex Adelman and Mrs. Olga Kazantseva, who provided assistance in many aspects of the book's preparation.

The authors will appreciate comments from readers to be used in further work.

1

INTRODUCTION TO DRILLING TECHNOLOGIES FOR OIL AND GAS IN RUSSIA AND THE FSU

Progress of Well Drilling Technology in the Second Half of the 19th Century and the Beginning of the 20th Century [1,2]

Russia and the Former Soviet Union (FSU) have taken the lead in the history of the development of world hydrocarbon resources more than once.

Historical facts indicate that in the eighth century, oil was produced from shallow wells in areas where it seeped to the surface in Azerbaijan. In 1729, oil-producing wells on the Apsheron peninsula were marked on a map. Oil production near the Ukhta River in the North-European part of Russia (now the Komi Republic) started in 1742. During the next century, in 1858, oil was produced on the Cheleken peninsula (Western Turkmeniya).

The official start of the oil industry is considered to be 1859—the first well drilled by entrepreneur E. Drake in Pennsylvania, USA, using the percussion drilling method. The well opened a new era of a wide-scale oil production. However, in 1848 in the Bibi-Eibat area of the Apsheron peninsula, a group led by F. A. Semyonov, an official from the Mining Directorate had drilled the first well using a mechanical drilling method. In 1864, Colonel A. N. Novoseltsev drilled the first productive well using the percussion method near Kudako village of the Kuban region.

The industrial boom in Russia began after the abolition of serfdom in 1861, which enabled tremendous growth in the oil industry.

The shallow well method and the drilling techniques developed later were widely used in the Russian mining industry long before the first oil wells were drilled. These methods were utilized to build wells for the production of salt and water and to explore mineral reserves, particularly coal. For example, four water wells, 36 to 189 m in depth, were drilled in 1831 in Odessa, Ukraine. Similar drilling was conducted in St. Petersburg, in the Crimea and in some other regions. All these wells were drilled using the percussion method.

Rapid growth of the oil industry resulted in the emergence of a number of top-notch professional mining engineers and toolpushers specializing in well drilling technology. Their cumulative experience was summarized in a number of books and publications. For example, one of the most outstanding books, *The Mining Art Course* by A. I. Uzatis, was published in 1849 in St. Petersburg. Surprisingly, the book forecasted many ideas that were used many years later in Russia and abroad. The author described percussion and rotary drilling techniques, well-casing technology, rod-tools, and rope drilling methods. Along with vertical wells, a number of directional wells were drilled, initially from inside the mines and later from the surface. One section of the book classified the wells as vertical, directional, and horizontal.[3]

For almost forty years, until the publication of *The Reference Book for Mining Engineers and Technicians* by Professor G. Y. Doroshenko in 1880, Uzatis's work remained the reference book for mining specialists, including oilmen. Later, between 1904 and 1911, one of the most prominent Russian mining engineers, I. N. Glushkov, published a four-volume classical work, *Well Drilling Manual.*

Based on these facts, it is clear that a large number of mining engineers were available and ready to manage oil and gas well drilling operations in Russia. After serfdom was abolished in Russia, the availability of hired labor also stimulated the development of the country's oil industry. In 1862, the Russian oil industry produced 5500 tons of oil, and by 1872, production had increased fivefold to 25,600 tons.

By 1880, extensive drilling experience allowed for rapid growth in oil production in the ensuing years. In 1885 for example, the main oil-producing region, Baku, had 500 producing wells, most of which had been drilled using the percussion rod drilling method. The rest were drilled with the rope (or Pennsylvania cable) technique. By 1899, the 944 percussion-drilled rigs belonged both to Russian and foreign operators in that area and included 881 percussion rod and 63 rope drilling rigs. These rigs drilled 174,300 m.

A number of outstanding mining engineers who worked during that period developed unique drilling methods and equipment to drill for both water and oil. G. D. Romanovsky, a mining engineer and leading scientist, was the first geologist to predict large oil deposits in the Mid Povolzhye (Volga River area) region in 1868. In 1866, he developed and applied the freefall drillstring bottom or cable end that automatically turned at a certain angle each time it hit the bottomhole. After the design of this tool had been improved by Dudin and Lents (one an engineer and the other a technician), it was widely used in percussion drilling. The development and use of the self-turning, freefall tool marked a significant step forward in drilling technology.

In the 1880s and 1890s, a series of new Molot drilling rigs with an improved design were developed by Mukhtarov and Lents.

S. G. Voyislav, a prominent mining engineer, made a major contribution to drilling technology. In 1885, he invented and built a special well borer for drilling large diameter upper-well sections. The tool used a reaming technology and allowed drillers to achieve a seven-times faster penetration rate and drill a large diameter hole as much as 22 m in depth.

In 1898, Voyislav and L. Kulesh invented and patented a rig with a diamond drilling system for drilling rock with various degrees of hardness by applying constant pressure. The rig design provided for automatic regulation of the penetration rate, depending on the hardness of the drilled rock. This rotary drilling rig had rotational speeds up to 7000 revolutions per minute (rpm). Voyislav significantly improved the diamond drillbit design, using a special method for diamond positioning and attachment in the bit body matrix, which achieved much better results when drilling. Thanks to these outstanding accomplishments, Voyislav is considered the originator of diamond drilling technology in Russia. The technology was also used in drilling exploratory oil wells.

In 1894, Voyislav was the first to use directional borehole drilling to drill a water well near the city of Bryansk. He was also the first to introduce the box joint to connect the steel drilling rod and eliminate the thread connections plane of weakness, which helped prevent a large number of rod connection failures.

The need to drill deeper wells in order to bring deep oil-bearing reservoirs into production prompted drilling engineers to develop rig power drives (such as steam, diesel, and electrical engines) to be used instead of a hand drive and horse traction. Data from the Baku region illustrates the progress of increasing well depths. In 1873, the average well depth was 22 m; in 1883, it was 59 m; and in 1893, it was 113.8 m.

Romanovsky pioneered the development of a drilling rig power drive by introducing the steam engine drive. In 1859, a water well in Podolsk, Moscow region was the first well drilled by a steam-driven rig. It was drilled at about the same time that Drake drilled his first well using a manually driven rig and wooden drill rods. The second steam-driven well was drilled between 1865 and 1869 near the Batrakov village, and the third well was drilled in the Crimea in 1877. This last well set a record depth of 750 m. Romanovsky's work was a breakthrough in the oil patch, primarily because of his achievements in mechanizing drilling operations.

In Baku, Grozny, and other regions of Russia, drillers were progressively moving towards the use of drilling rigs with steam engine drives which, by the end of the 19th century, completely replaced the manual and horse drives. Still, the steam engines at that time had a low efficiency factor (2–3%) and consumed a significant amount of fuel (up to 13% of the produced oil volume). Therefore, by the early 20th century, oil producers were using internal combustion engines and electric motors. Along with increased in-well depth, drillers began using rotary instead of percussive drilling techniques.

In 1902, a rotary rig equipped with a drilling mud circulating system drilled its first well in Russia near Grozny. The well depth was 345 m. In 1908, the company, Shpis, made a second attempt to drill several wells using a rotary drilling rig but eventually refused to use these rigs further. In 1906, a company owned by the Nobel brothers in cooperation with the Nafta Company made their first (and unsuccessful) attempt to drill a well with a rotary rig. The two wells were drilled to 520 m and 720 m and had deviated boreholes up to 30–40°. The attempts failed because of the poorly designed primitive rotary drilling rig and lack of experience in applying the new drilling technique. In 1911, oil producer Gaber managed to achieve a positive result when two of the eight wells he started were drilled to total depth of the well (TD) and put into production. Unlike the progress of the percussion drilling method in the 19th century, development of rotary drilling in Russia in the early 20th century was rather slow. For example, of the 14 companies working in the Surakhansky region near Baku, only one used a rotary drilling rig. One of the first companies started by the Nobel brothers exclusively used the percussion drilling method. Before the oil industry was nationalized in 1920, this company had drilled only 12 wells with rotary drilling rigs.

The most probable explanation of this fact is a series of dramatic social and political events, such as Russia's defeat in the war with Japan (from 1904 to 1905), the Revolutions in 1905 and 1917, and the World War I from 1914 to 1918. All these events were alarming for oil business people and provided little

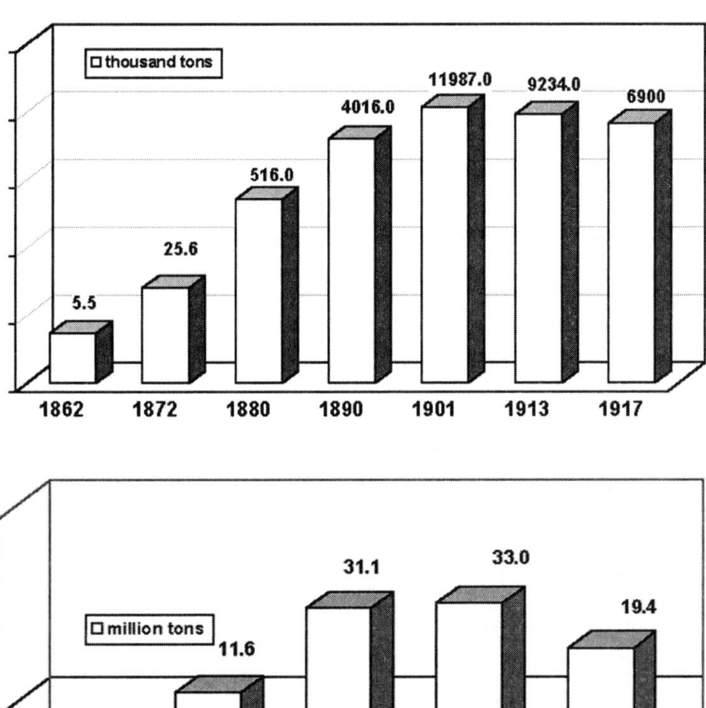

Fig. 1–1 Oil production in Russia and in U.S.S.R. until the end of WWII

encouragement for investment in the improvement of drilling equipment and technology. The development of capitalism in Russia, which had produced extremely high results in the second half of the 19th century, slowed rapidly.

Introduction of the well-cementing method was among the achievements of the drilling technology at this time. Romanovsky was among the first to apply this technology. While drilling a water well in 1859, he ran a special cylindrical container with cement slurry in the hole. After the cement filled the borehole and shrank, a new hole was drilled through the cement column. Thus, the new

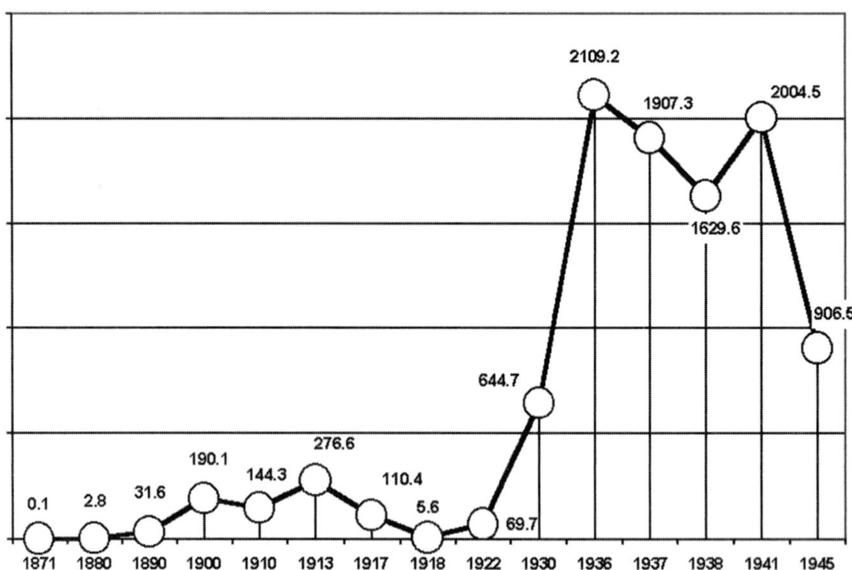

Fig. 1–2 Oil and gas drilling volumes in Russia and in the U.S.S.R. before the end of WWII (thousand meters)

borehole walls were stabilized by the cement sheath and eliminated the need to run a casing string. A casing cementing method was patented by the Russian engineer Bogushevsky and was similar to the method developed by Perkins. This method was first used in Russia in 1906.

According to a law adopted in 1908, oil producers were obliged to cement casing strings in wells to isolate oil-bearing horizons from water zones. However, the requirements of this law were ignored for several years, which resulted in dramatic flooding of the highly productive horizons and the loss of significant oil reserves. Later, the need for cement jobs was recognized, and cementing was performed in all drilled wells.

The Russian drilling boom previously described ensured steady oil production growth. Diagrams in Figures 1–1 and 1–2 indicate that 276,500 m were drilled in 1913, compared with 100 m and 190,100 m drilled in 1871 and 1900 respectively. Corresponding oil production increased from 25,600 tons in 1872 to 12 million tons in 1901. This was the highest oil production level in Russia before WWI, and Russia became the number one world oil producer in the beginning of the 20th century. In 1905, world production was 26 million tons. However, by 1917, the oil production level in Russia was down to 6.9 million tons a year.

Milestones of Drilling Technology Development in the FSU after Nationalization of the Oil Industry (1920–1945)

During the October Revolution in 1917 and the Civil War in Russia, drilling activity and oil production continued to decline. In 1918, only 5600 m were drilled. Oil production in 1922 amounted to 4.7 million tons. The Russian oil industry was set back about 30 years. After the oil industry was nationalized in 1920, an effort was made to revive the ruined industry and raise drilling activity and oil production. As had been done before, special significance was given to the Baku region, because it was the main producer and supplier of oil and oil products.

Azneft, the production association, was formed. The company managed regional oil production divisions that were formed by bringing together the oil wells, drilling rigs, and other industrial facilities expropriated from their former owners. A similar management scheme was adopted in other regions such as Grozny and Krasnodar. The state financed drilling and production operations.

Despite the country's devastation and economic woes, the government understood the strategic role of the oil industry in the country's economy revival and managed to allocate financial resources for its development. Before the 1917 Revolution, western companies provided a significant part of the total investment in the Russian oil sector (up to 56% or 460 million rubles). The British (37% of the total investment) and French (13%) companies owned 60% of produced oil and took 75% of the oil products market in Russia.[4] Obviously, the state had to find significant financial resources to invest in the oil industry. In the early 1920s, money was used to rehabilitate the Russian oil industry. Development and improvement of drilling technology and oil recovery methods were among the high priority objectives.

These objectives were achieved by replacing old methods and technologies with newer and advanced ones; for example, rotary drilling was used instead of percussion drilling, downhole air injection and gas lift well operation were used instead of bailing. During those 10–15 years when Russia was struggling through dramatic events, the United States and other oil-producing countries had succeeded in developing these new techniques.

A. P. Serebrovskyi, a prominent Russian statesman with a mining engineering background, was appointed Director of the Azneft Company. He was also Chairman of the Board of the All-Russia Oil Syndicate and Deputy Chairman of the All-Russia National Economic Council (VSNKh). He was a man of marked initiative and intellect and managed Azneft from 1920 to 1926. He sought out highly qualified and dynamic specialists and placed them in charge of oilfield production facilities, drilling companies, and auxiliary service enterprises.

A great deal of effort on the part of Azneft personnel yielded positive results. By October 1920, 71 drilling rigs were assembled and put into operation in Baku, including 62 percussion and 9 rotary rigs. Still, manufacture of percussion drilling rigs had not been fully resumed, and rotary drilling rigs were not built in the country at that time. Azneft decided to import rotary drilling rigs with all the necessary tools and materials from the United States and to contract qualified drilling consultants. The development program also called for the import of production equipment such as submersible pumps.

In 1921 and 1922, Serebrovskyi, much trusted by V. I. Lenin, Chairman of the National Commissars Council (Sovnarkom), received the required financing and successfully completed the modernization program. The fast and successful completion of this critically important work did much to improve the situation in the oil industry of the Azerbaijan Republic. After Serebrovskyi visited the United States, he wrote a book, *The American Oil and Gas Industry.* It was the first detailed coverage of the U.S. oil industry development. From 1923 to 1924, the proportion of the operating rotary rigs increased rapidly and resulted in faster penetration rates and more drilling.

Figure 1–3 shows the proportion of wells drilled by rotary and percussion drilling rigs from 1913 to 1935. The diagrams clearly indicate that by the end of this period, all wells were drilled using the rotary technique, which allowed an increase in working efficiency as well as overall drilling rates (Fig. 1–4) [5].

Application of the imported drilling rigs and other equipment gave an impetus to the development and improvement of drilling technology. Several innovations were introduced. One was the blade bit RKh (fishtail) with the heat-treated or hard faced blades. Simpler well designs resulted in a lower number of casing strings and a longer open-hole section drilled below the previous string shoe. The latter innovative technique enabled the significant reduction of steel consumption per drilled meter. Thus, steel consumption in the Azneft Company went down from 320 kg/m in 1926 and 1927 to 100 kg/m in 1931. The corresponding figures for Grozneft were 210 kg/m and 60 kg/m.

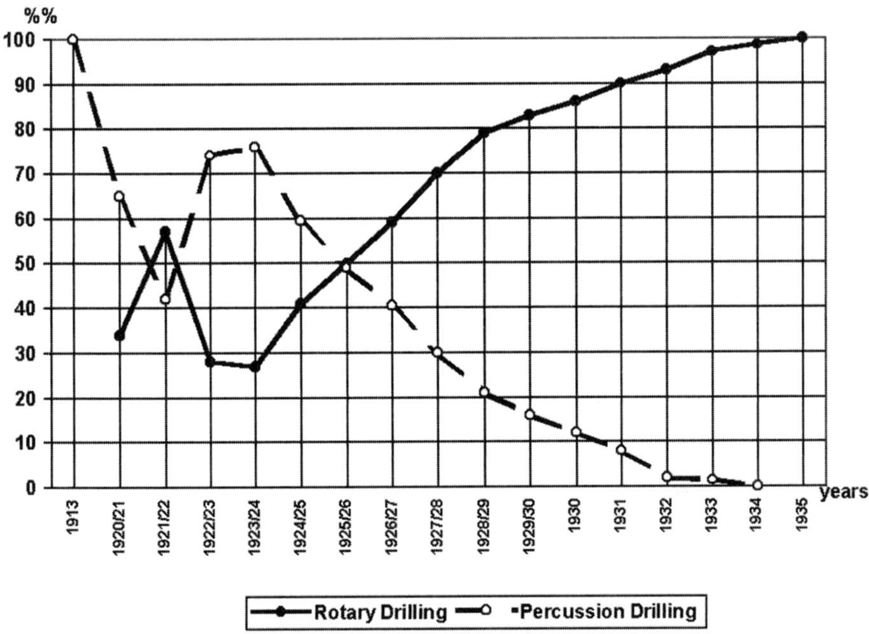

Fig. 1–3 Utilization of percussion and rotary drilling methods in Russia and FSU in 1913–1935

The industry began using the well cementing method developed by Perkins rather than the annulus cementing method or the technique that squeezed the casing shoe into the clay rock. Drillers applied chemicals to accelerate thickening of the cement slurry, which reduced the waiting-on-cement (WOC) time in 1928 from 14 to 7 days. After a few years the WOC time was reduced to 72 hours. In 1939, wooden derricks were replaced with steel structures. Their design was continuously improved, and they became much lighter. Steel bracing members of the derrick were introduced in the early 1940s. Corner supports and derrick girth were made of drillpipe. Rigging-up methods for the derrick were also improved. Derricks were assembled from the top down using the Kershenbaum technique. Derrick-bracing members were moved around on special carriages.

In the meantime, the industry began using new higher capacity rigs with three and four drive gears instead of the old two-gear design.

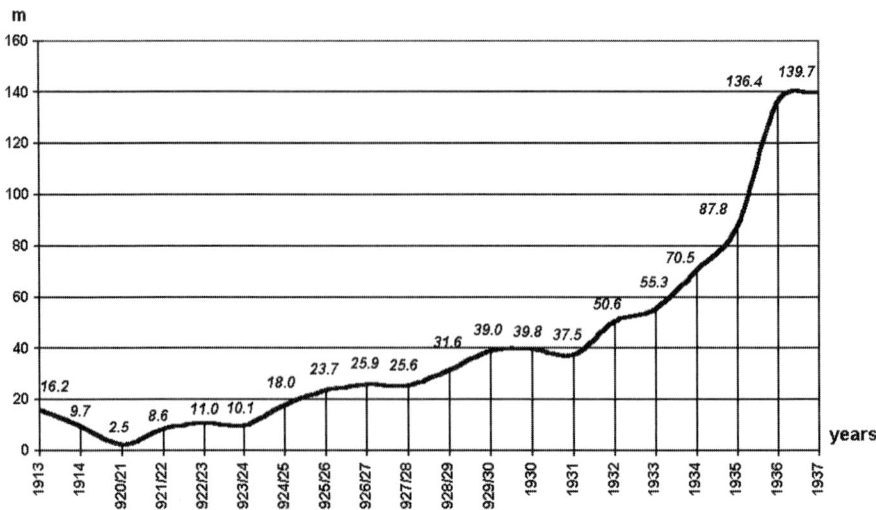

Fig. 1–4a Performance of drilling operations in 1913–1937; penetration per one worker in meters

In the late 1920s and early 1930s, Russian plants started manufacturing various drilling equipment, such as drillbits and the materials and tools used in drilling. The manufacturers mostly used American designs for these items. The plant in Baku, named after Lieutenant Schmidt began manufacturing 60-, 100-, and 150-ton two-speed and four-speed drawworks. The components of the block and tackle system were made at other mills in Baku. The Krasnyi Molot plant in Grozny produced mud pumps. In 1933, domestic manufacturers delivered most of the two-speed drawworks, open-type rotary tables, and mud pumps. Until the end of the 1940s, the development of the oil-related manufacturing industry and drilling technology in Russia followed the path of the U.S. oil industry, lagging behind slightly. The related imports continuously declined and practically ceased in the 1940s. At the same time, measures were implemented that aimed at revival and development of the oil industry in the FSU. These measures promoted the rapid growth of drilling and oil production rates. Information in Figures 1–1 and 1–2 shows the maximum levels of drilled footage (meterage)—more than 2,000,000 m per year—were achieved in 1936 and 1941. Oil production reached its maximum level of 33.0 million tons in 1941.

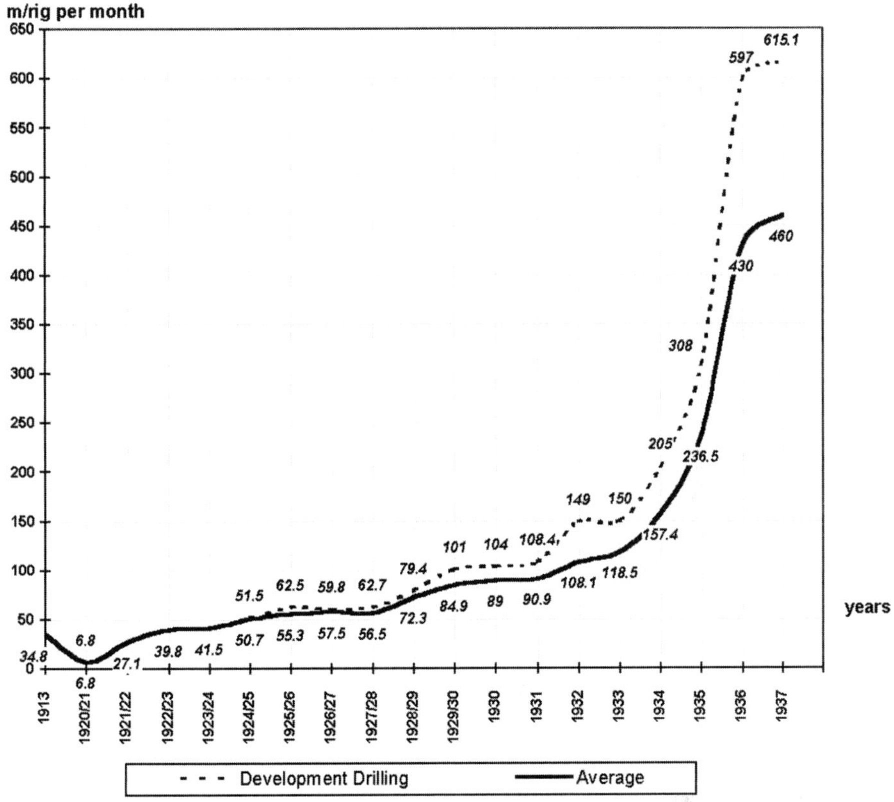

Fig. 1–4b Performance of drilling operations in 1913–1937; overall drilling rate (m/rig per month)

The drilling slowdown in 1937 and 1938 was due to the J. Stalin's repression. For every case of drillpipe failure, toolpushers and drillers were arrested. To avoid such failures, they had to lower the bit weight and drillstring rotational speed, which resulted in a reduced penetration rate. The penetration rate caught up with previous levels only in 1941. Table 1–1 presents detailed information for every year and region. [6]

TABLE 1–1
Drilling Volumes in Russia and FSU by Regions in 1871—1941

Year	Azerbaijan and Georgia	Grozny	Krasnodar Region	Emba Regions (Kazakhstan)	Turkmeniya	Middle Asia	Sakhalin	Urals-Volga Province	Total
1871	0.1	–	–	–	–	–	–	–	0.1
1872	-	–	–	–	–	–	–	–	–
1873	2.7	–	–	–	–	–	–	–	2.7
1874	3.0	–	–	–	–	–	–	–	3.0
1875	1.6	–	–	–	–	–	–	–	1.6
1876	2.9	–	–	–	–	–	–	–	2.9
1877	1.7	–	–	–	–	–	–	–	1.7
1878	5.3	–	–	–	–	–	–	–	5.3
1879	7.1	–	–	–	–	–	–	–	7.1
1880	2.8	–	–	–	–	–	–	–	2.8
1881	4.8	–	–	–	–	–	–	–	4.8
1882	6.5	–	–	–	–	–	–	–	6.5
1883	5.0	–	–	–	–	–	–	–	5.0
1884	11.0	–	–	–	–	–	–	–	11.0
1885	12.6	–	–	–	–	–	–	–	12.6
1886	14.1	–	–	–	–	–	–	–	14.1
1887	20.6	–	–	–	–	–	–	–	20.6
1888	17.9	–	–	–	–	–	–	–	17.9
1889	29.0	–	–	–	–	–	–	–	29.0
1890	31.6	–	–	–	–	–	–	–	31.6
1891	42.7	–	–	–	–	–	–	–	42.7
1892	25.0	–	–	–	–	–	–	–	25.0
1893	23.3	0.1	–	–	–	–	–	–	23.4
1894	25.4	0.1	–	–	–	–	–	–	25.5
1895	43.5	0.5	–	–	–	–	–	–	44.0
1896	59.8	5.8	–	–	–	–	–	–	65.6
1897	84.8	–	–	–	–	–	–	–	84.8
1898	122.4	7.7	–	–	–	–	–	–	130.1
1899	174.1	10.3	–	–	–	–	–	–	184.4
1900	176.9	13.2	–	–	–	–	–	–	190.1
1901	161.2	16.2	–	–	–	–	–	–	177.4
1902	85.5	9.3	–	–	–	–	–	–	94.8
1903	104.7	13.6	–	–	–	–	–	–	118.3
1904	132.7	11.4	–	–	–	–	–	–	144.1
1905	75.9	10.9	–	–	–	–	–	–	86.8
1906	102.4	11.7	–	–	–	–	–	–	114.1
1907	130.3	16.1	–	–	–	–	–	–	146.4

1908	130.0	19.3	–	–	–	–	–	–	149.3
1909	116.0	25.2	–	–	–	–	–	–	141.2
1910	108.0	26.7	9.6	–	–	–	–	–	144.3
1911	104.0	21.8	25.5	1.4	–	–	–	–	152.7
1912	133.0	36.2	22.3	3.4	–	–	–	–	194.9
1913	171.8	62.9	30.9	9.2	–	1.8–	–	–	276.6
1914	142.5	85.0	13.0	9.6	–	–	–	–	250.1
1915	126.0	54.4	11.6	7.8	–	–	–	–	199.8
1916	118.5	48.6	5.5	8.7	–	–	–	–	181.3
1917	69.5	38.7	–	2.2	–	–	–	–	110.4
1918	5.4	–	–	0.2	–	–	–	–	5.6
1919	13.2	–	–	–	–	–	–	–	13.2
1920	6.2	1.3	–	–	–	–	–	–	7.5
1920–1921	3.4	1.8	0.5	0.2	–	–	–	–	5.9
1921–1922	15.0	3.4	0.5	0.2	–	–	–	–	19.1
1922–1923	50.5	16.8	1.9	0.5	–	–	–	–	69.7
1923–1924	77.6	40.5	3.5	1.6	–	–	–	–	123.2
1924–1925	118.8	57.0	2.4	3.9	–	0.1	–	–	182.2
1925–1926	200.8	74.9	5.0	5.7	–	1.5	–	–	287.9
1926–1927	253.1	104.8	8.7	12.2	0.4	2.5	–	–	381.7
1927–1928	260.6	74.7	11.7	13.1	0.3	1.7	–	–	362.1
1928–1929	320.5	87.8	13.5	18.0	0.3	3.2	2.7	–	446.0
1929–1930	404.8	106.6	18.5	29.6	2.2	6.8	5.4	11.7	585.6
1930*	117.9	29.0	6.6	10.5	0.4	2.3	0.3	3.5	170.5
1931	455.3	122.0	38.5	49.0	2.0	13.9	8.1	18.4	707.2
1932	490.7	101.1	49.0	44.6	2.9	18.5	17.6	20.3	744.7
1933	540.6	137.5	47.2	36.3	2.5	19.4	18.3	33.7	835.5
1934	807.2	243.9	65.9	47.0	5.2	24.4	20.8	39.2	1,253.6
1935	969.6	266.0	103.5	48.0	7.4	19.4	22.6	69.1	1,505.6
1936	1,404.91	270.6	151.0	56.7	18.7	32.6	39.4	86.5	2,060.4
1937	1,202.1	219.1	124.2	64.4	24.4	43.3	40.6	119.4	1,837.5
1938	954.9	215.7	124.2	74.7	20.3	46.6	25.2	149.1	1,610.7
1939	923.1	221.9	124.5	77.7	19.9	43.9	18.4	186.5	1,615.9
1940	1,002.7	287.4	120.1	87.2	16.5	39.6	14.4	223.2	1,791.1
1941	929.8	349.3	130.4	120.1	42.3	38.4	19.7	254.5	1,884.5

**Special quarter was added to start recording from the January 1 next year.*

By 1945, an even faster decline in oil production took place during World War II. Despite the difficult times, certain positive features stood out among the conventional world oil industry practices. These related to the development of drilling equipment and technology in Russia and included the following aspects.

(1) The introduction drilling rigs driven by electric motors for greater efficiency and power saving. By 1931, drillers in Baku had replaced all the steam engines on drilling rigs with electric motor drives. In Grozny, 88.3% of the drilling rigs used an electric motor drive.

(2) The large scale use of electric motors allowed the application of independent rotary drives (IRD), which resulted in higher bit rpm and more efficient control of drillbits rotational speed. In the late 1930s and early 1940s, the rotary speed in upper intervals of the well was as high as 180–200 rpm and resulted in significantly increased penetration rates.

(3) In 1924 and 1925, an artificial onshore field was built in Bibi-Eibat near Baku. Well No. 61 was drilled on the island to a depth of 460 m and flowed with oil. This discovery triggered work on expanding the onshore area into the sea by filling a large littoral area in Ilyich Bay with soil brought from other areas. The new oilfield increased the oil reserves and commercial oil production.

(4) Starting in 1922, downhole drilling motors with hydraulic and electric drives were developed intensively. By 1941, efficient hydraulic turbodrills were built and used to drill numerous pilot deep wells on the Apsheron Peninsula. In addition, the first electric downhole motor (EDM) was developed and used to drill a well to a depth of 1500 m. The development and wide use of downhole motors (DHM) reflected the evolution of the Russian drilling technology and resulted in the formation of the basic features of domestic drilling industry.

(5) The turbodrills were used to drill experimental directional wells in 1939 in Grozny and in 1941 in Baku. They proved to be more efficient for directional applications compared to rotary drilling combined with whipstocks.

(6) During World War II in 1943, the Russian drilling industry pioneered the use of turbodrills for cluster directional drilling while developing the Krasnokamskoye oilfield near the City of Krasnokamsk in the Perm region of the North Urals.

Principal Stages of Drilling Technology Development in the FSU and Drilling Operations Management in the Postwar Period

Development and implementation of turbodrilling for vertical and directional single and cluster wells

After nationalization of the oil industry from 1922 to 1941, engineers and researchers consistently developed reliable designs of hydraulic downhole motors (HDHM) and EDMs. By 1941, these tools successfully drilled several wells in the Baku region. Work was conducted on a commercial basis by the specialists of the Experimental Turbine Drilling Bureau (EKTB) of the Ordzhonikidzeneft Company in Surakhany, Baku region, starting in 1934.

The EKTB was reorganized in 1939 and placed under the authority of the U.S.S.R. Oil Ministry (NARKOMNEFT) in Moscow, which provided research and development (R & D) work at an appropriate level and ensured significant contributions toward its success. The EKTB formed in Baku indicated the emphasis of the Russian Oil Ministry and the government on the importance of turbine drilling development.[7,8] By 1941, the industry was using a large number of these efficiently operating turbodrills with designs that featured multi-stage turbines and rubber-metal axial and radial plain bearings, as well as all the tools to service this equipment.

After the war began, EKTB was relocated to Krasnokamsk, the Perm region, including all personnel and equipment. A large oilfield with oil deposits located in a carboniferous zone of Paleozoic at depths of 950–1000 m had been discovered in this region. A company established earlier in that region had been drilling wells using rotary drilling rigs. Afterwards, the joint Turbine Drilling Bureau was founded. The bureau became the first drilling company that used the turbodrilling method to drill exploration and development wells. Application of the new method allowed considerable improvement of vertical drilling results.

However, a significant breakthrough in the development of drilling operations did not take place until 1943 when turbodrilling technology was used for directional single and cluster wells.[9] The improvement in drilling efficiency was so great that by the end of 1943 and the beginning of 1944, the cluster pattern of development wells was used for the entire field.[10,11,12] In wartime conditions, this outstanding event marked a milestone of development in the FSU drilling industry. Vertical and directional turbodrilling as well as cluster well drilling became widely used in commercial drilling applications for oil and gas in all regions of the country.

By the late 1950s, 80% of the wells in the FSU were drilled with turbine drilling technology. To aid in implementing this technology, a special Department of Turbine Drilling was formed in 1942 at the NARKOMNEFT. R. A. Ioannesyan and M. G. Gusman, the inventors of the turbodrill, headed this department. The department's objective was to implement turbodrill technology for vertical and directional wells in various regions of the country.

In 1944, special Turbine Drilling Bureaus were formed in many cities and regions of the FSU, such as Baku, Grozny, Kuibyshev, and the Republics of Dagestan, Bashkiriya, and Tatariya. The bureaus aimed at improving and implementing turbodrill technology in these regions. The bureaus used the facilities and qualified drilling personnel of the rotary drilling departments. Turbodrill specialists from the EKTB in Baku and the Bureau of Turbine Drilling in Krasnokamsk were appointed to management positions in the new bureaus. From 1944 to 1946, the Department of Turbine Drilling arranged and coordinated efforts to supply the required drilling equipment to the new companies. The equipment included turbodrills, imported mud pumps, high-pressure mud hoses, various auxiliary tools, and spare parts. The Myasnikov plant in Baku carried out full-scale turbodrill manufacturing. In 1941, the plant was evacuated from Baku to Pavlovsk, a town in the Perm region.

Besides the technological and organizational problems encountered during the wide-scale deployment of turbodrilling methods, other factors affected the development of this technology in Russia. Opponents of the method cited the experience of drilling industries in the United States and other countries, saying that the Russian drilling industry had followed the wrong trend. According to them, this trend would result in lowering the standards of the metallurgical and other industries with respect to manufacturing high-strength drillpipe and drillcollars (DRC), reliable rotary units, swivels, and other equipment used on a rotary rig.

In fairness, we should say that in the second half of the 1940s and in the 1950s, some of the skeptics' critical remarks were validated, because, in some cases, rotary drilling had more advantages than turbodrilling. As happened in many other cases under the Soviet regime, when influenced by the general euphoria of the success of a new technology, the officials imposed the use of turbodrilling in the southern regions of the country. For deep wells (3000–3500 m TD) in the regions of Baku, Krasnodar, and Stavropol, this was certainly a mistake. Diamond bits, low-speed high-torque turbodrills, and positive displacement motors (PDMs) were not available yet, and rotary drilling proved to be more efficient in the lower intervals of these wells.

The geological sections of the southern oil-producing region featured relatively young Neozoic and Mesozoic deposits primarily composed of soft and medium sandy argillaceous rock. The turbines available at that time were not capable of providing the high torque and low rpm required to efficiently destroy rock with roller-cone or blade-type drag bits. The bit rotational speed was too high (600–800 rpm), and the torque level was limited, which resulted in reduced penetration per bit due to low-bit durability and shorter bit-on-bottom time. Therefore, by the late 1950s, in these regions as well as in other southern regions, turbodrilling was used primarily in directional wells with depths of 2000–2500 m, especially offshore Caspian near Baku and Dagestan. Later in the 1960s and 1970s, DHMs became more widely used, thanks to the introduction of the two- and three-section turbodrills and PDMs with low rotational speed and high torque as well as high-speed diamond bits.

Further experience proved the expedience of a high priority development plan for DHM drilling technology, which could be explained as follows.

The state of the postwar metallurgical and machine building industries did not allow for the wide-scale production or use of rotary drilling rigs. The main industrial potential in metallurgy and mechanical engineering were concentrated in the military-industrial establishment (MIE) of the FSU. For a few years, MIE was applied to the oil industry (after 1945), but this was terminated by the beginning of the Cold War. Only a few oilfield equipment items were not affected by the industry conversion to defense-oriented production.

The Uralmash plant in Sverdlovsk (now Yekaterinburg) produced certain units of heavy duty drilling rigs. The Barrikady plant in Volgagrad manufactured light-weight, compact drilling rigs and DRCs. The Motovilikha plant in Perm produced small batches of turbodrills. The plants, operated under the authority of the Ministry of Oil and Petrochemical Engineering, supplied most of the tools and equipment to the drilling companies.

At the same time, production of the tri-cone roller bits at the Dzerzhinsky plant in Baku and the Verkhne-Sergiyevsky plant in the Urals faced some serious problems. The quality of the Russian drillbits was much lower, compared to the bits of the leading U.S. bit manufacturers such as Hughes, Reed, and Security. However, the limited number of trips and related auxiliary handling operations for 2000–2500 m turbodrilled wells suggested that rate of penetration (ROP) was the main factor affecting the overall drilling speed and well cost. The penetration rate of turbo-drilling is normally higher than the rotary drilling penetration rate.

In the 1950s and 1960s, the issue of turbodrilling efficiency compared to true American rotary drilling was discussed more than once. Tests related to the issue were performed with the participation of U.S. drilling experts, such as the test for drilling vertical wells in the Tatariya and Bashkiriya regions. Unlike the test for turbodrilling, most of the tests for rotary drilling used western bits. Despite the low quality of the drillbits made in Russia, in most cases the turbodrilling test results were on the same level with rotary test results or even better. After the tests, the drilling experts, including the U.S. specialists, concluded it was not advisable to use rotary drilling technology instead of turbodrilling in these regions.

The rapid growth of directional and cluster well drilling activity confirmed the advantages of DHM drilling technology in most of the oil- and gas-producing regions. Most of the directional cluster wells were drilled in onshore oil and gas fields. This can be explained by specific geographic conditions in the FSU, such as remote oil and gas fields, poorly developed field infrastructure, and large territories of tundra, marshes, and permafrost (Fig. 1–5). In these conditions, the ability to drill three, four, or more wells from one location reduced the number of drilling locations, expedited field development, and lowered operating cost. The method proved to be particularly efficient in Western Siberia where most of the factors that complicated the drilling process were found.

From the second half of the 1960s until today, practically all the Siberian oilfields were developed by cluster well drilling using HDHM. This was the key factor for the tremendous growth of the drilling and oil production volumes in the FSU during those decades, which made the FSU the world's leading producer of hydrocarbon products. The data on oil and gas production is presented in Figures 1–6 and 1–7, and the charts in Figure 1–8 describe the drilling activity in the FSU.

Most prominent was the implementation of turbodrilling at the end of the 1950s when it grew to 80% of total drilling volume. This happened simultaneously with development in Western Siberia. The data available clearly indicates a correlation between the rapid growth of drilling volumes in Western Siberia and the increase in oil production in the FSU (Fig. 1–9 and 1–10).

The new fields in the FSU provinces that were developed after WWII—Volga-Urals and Western Siberia—could be categorized as normal pressure, while the fields in provinces such as Azerbaijan, North Caucases, and Turkmeniya required drilling under high-temperature and high-pressure (HTHP) conditions. The discoveries made recently in West Kazakhstan and the North Caspian areas were in the HTHP category with additional hydrogen sulfide (H_2S) conditions. The average well depth in a conventional field exceeded 3000 m, and HTHP wells exceeded 4500 m in depth.

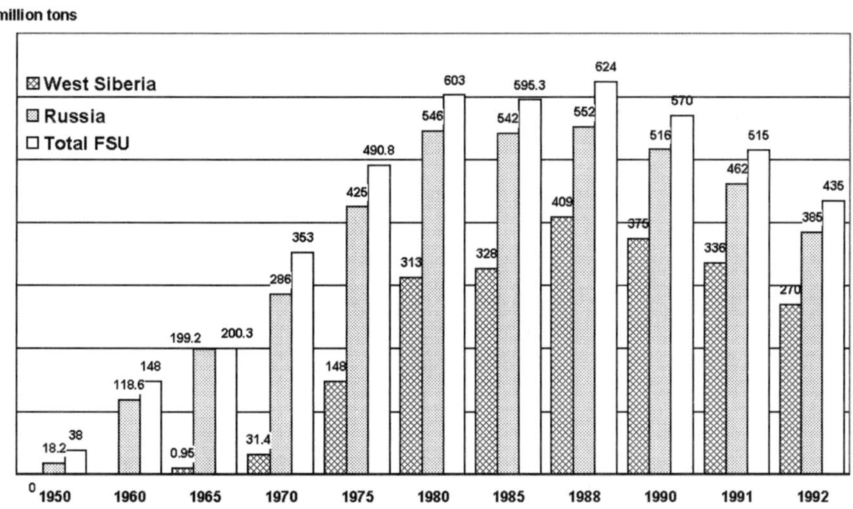

Fig. 1–6 Oil production in FSU after WWII (million tons)

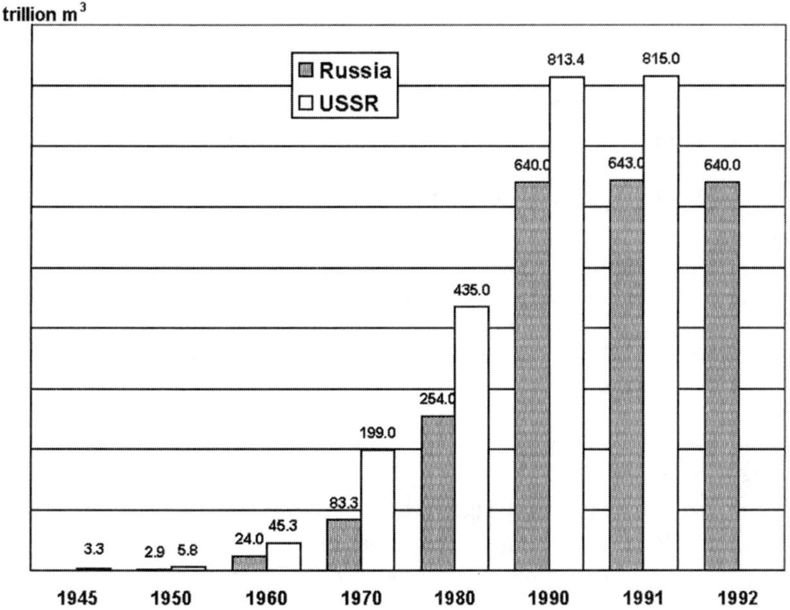

Fig. 1–7 Gas production in FSU (trillion m³)

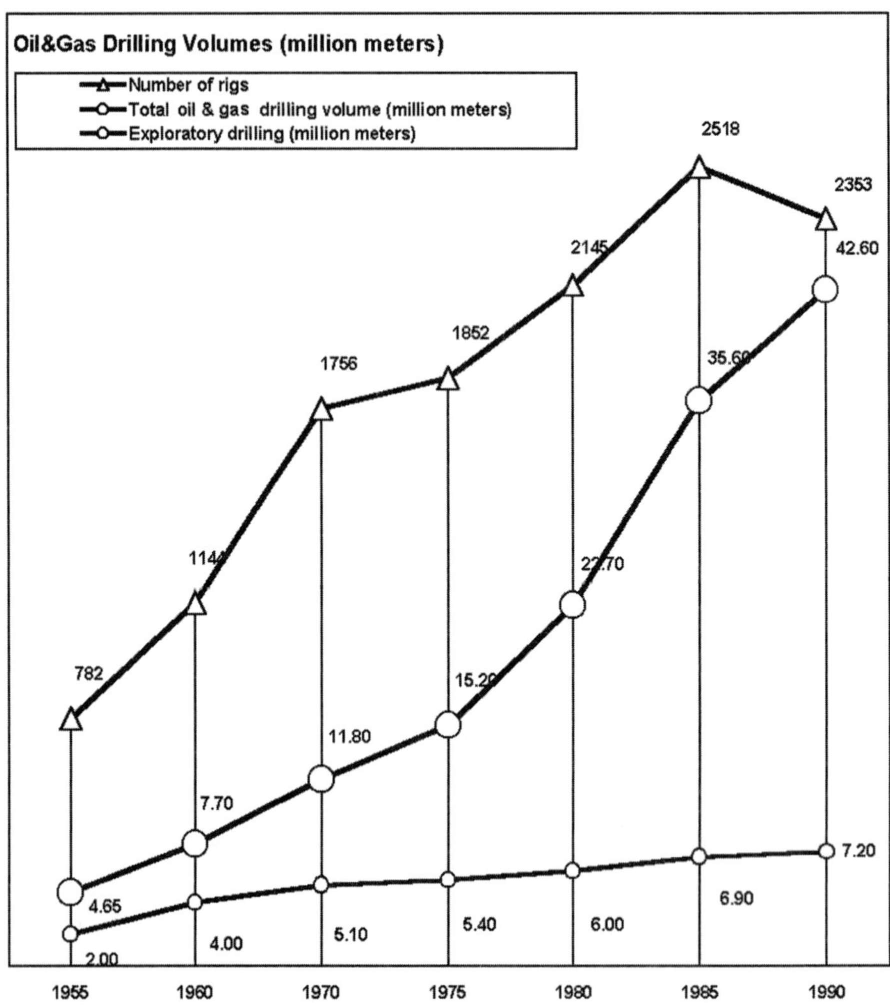

Fig. 1–8a Drilling activity after WWII: drilling volumes and rig count

The advantages of the widely used DHM technology were so obvious that the technology itself managed to survive through another strong rotary drilling campaign. This campaign was triggered by the introduction in the United States of the more durable tri-cone tungsten carbide insert (TCI) bit design with sealed bearings. The bit found supporters among the officials of the FSU Oil Ministry.

Introduction of the new bits was a breakthrough in the improvement of rotary drilling and resulted in the higher efficiency of deep well drilling because of the

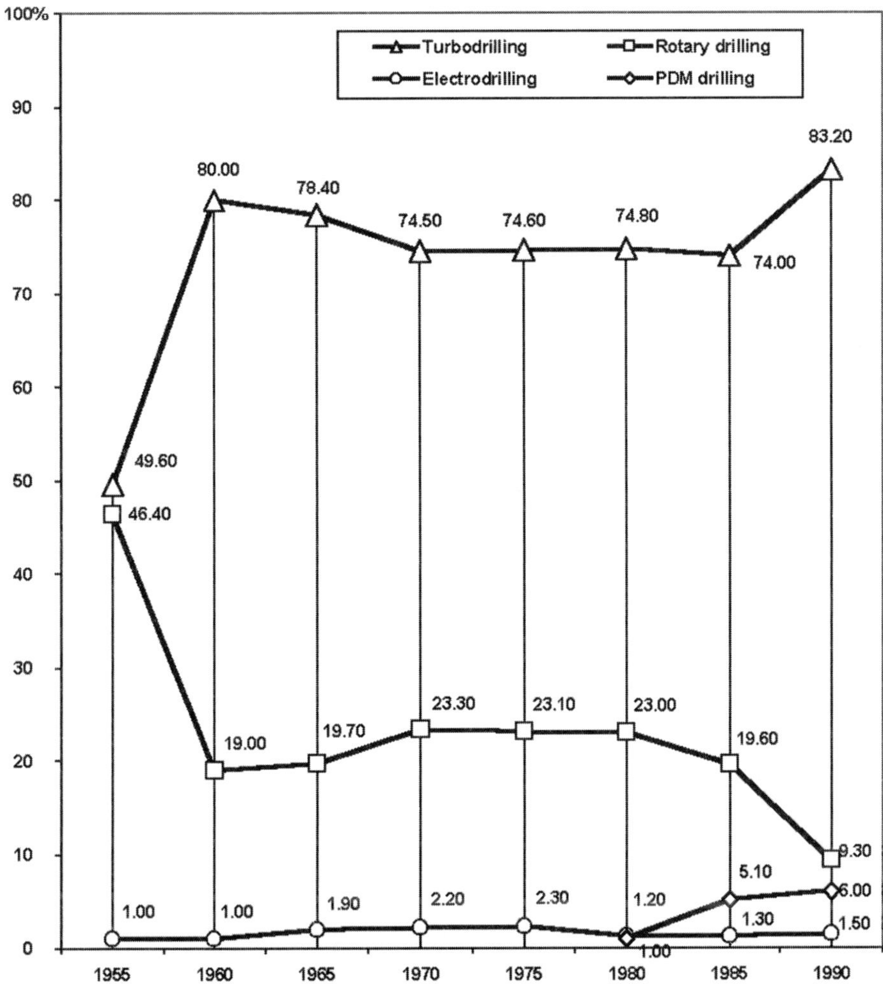

Fig. 1–8b Drilling activity after WWII: utilization of different drilling methods

reduced trip time. However, this innovation failed to improve penetration rates and even slowed down the overall drilling speed in some cases because drillbits with sealed bearings required lower drillstring rotation speed. This proved the dominating effect of the ROP on the overall drilling parameters for the wells with TDs of up to 2500 m, when higher ROP achieved by higher drillbit rpm. The higher rpm in turn required application of appropriate DHM, not rotary method.

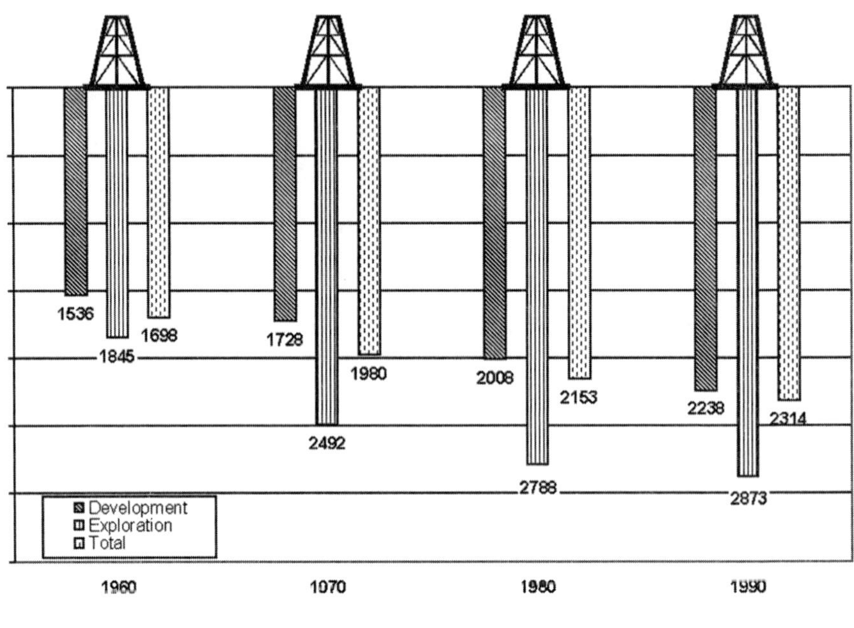

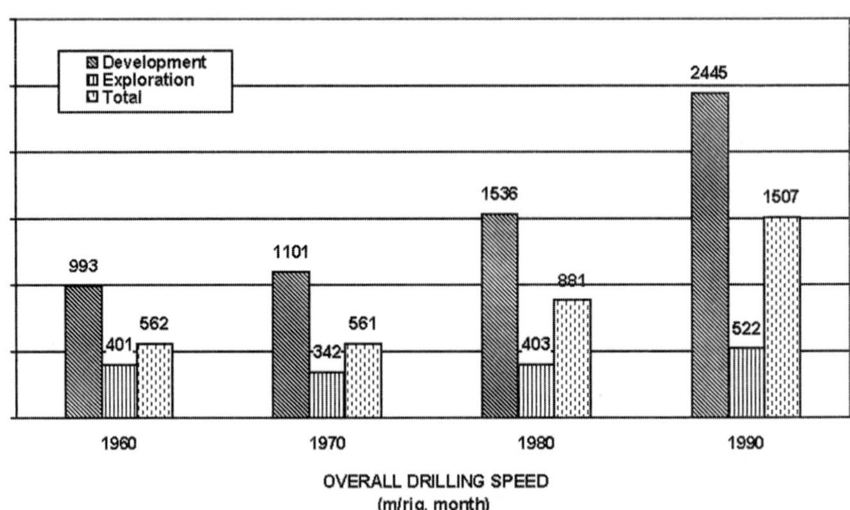

Fig. 1–8c Drilling activity after WWII: average wells depth and overall drilling rate

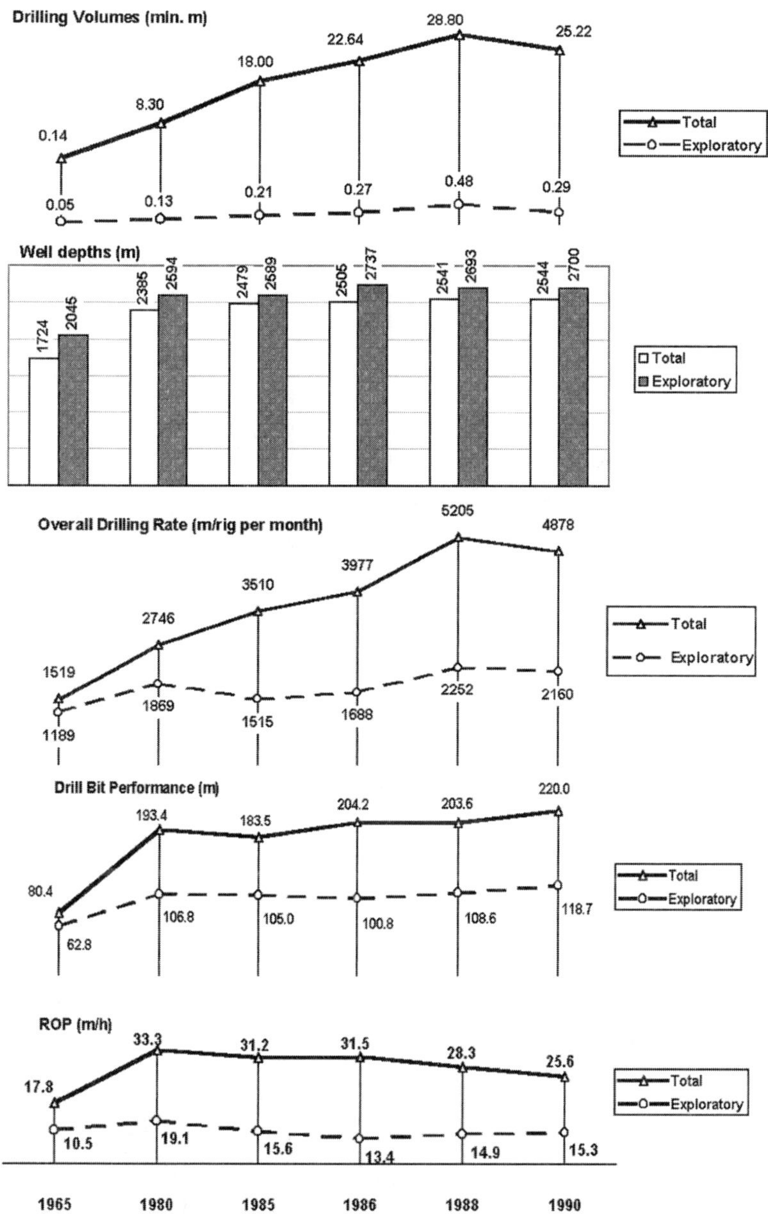

Fig. 1–9 Drilling operational results for West Siberia

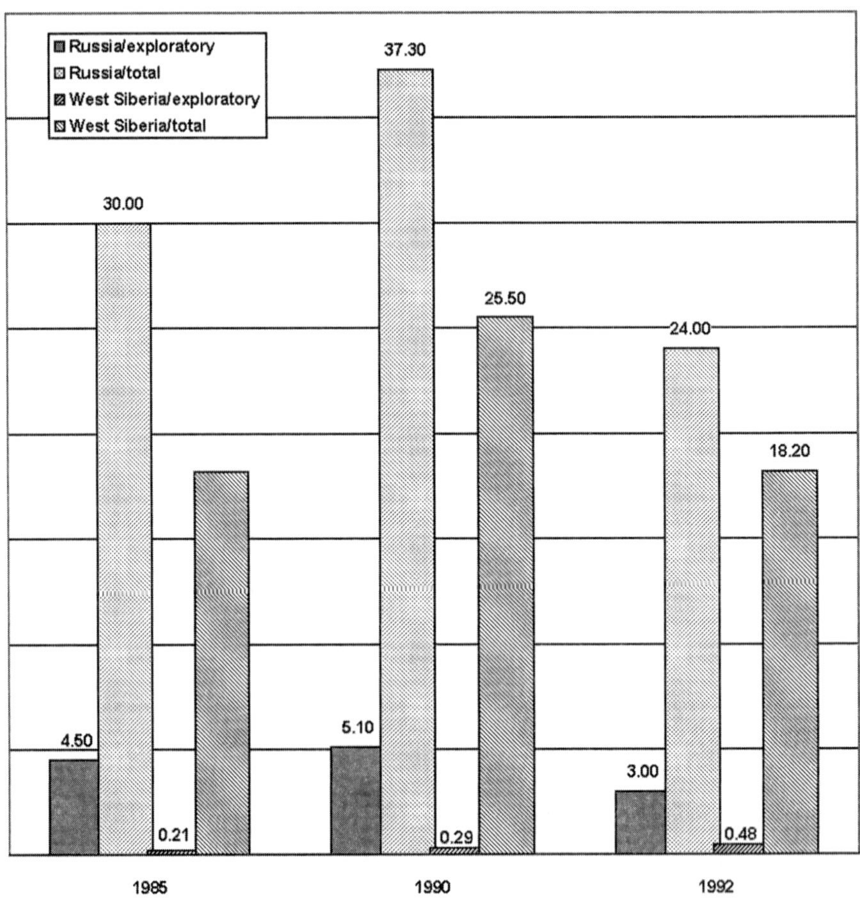

Fig. 1–10 Drilling volumes comparison in Russia and West Siberia (million meters)

Attempts of the drilling companies in some regions, including Western Siberia, to return to using the rotary drilling method failed completely. In 1979 and 1980, engineers from the All-Union Scientific and Research Institute of Drilling Technology (VNIIBT) conducted a special study to compare the results of turbodrilling and rotary drilling using the sealed bearing bit. The study convincingly proved the validity of the functional dependency of overall drilling parameters on ROP. Chapter 3 presents details of this study.

The soundness of using DHMs, both turbodrills and PDMs, in the FSU was supported by their increasing use in the West due to the rapid development of offshore drilling. Most offshore wells were drilled from one location and had directional boreholes. The potential of drilling deep wells using low rotary speed roller-cone bits was practically spent, since further increases in penetration per bit run failed to produce tangible improvements. The ROP level set the limit for further increase of a drilling speed that requires higher bit rpm. Use of the low speed, high-torque multi-lobe PDM, developed in the 1970s, improved the situation but not significantly. The growing application of polycrystalline diamond compacts (PDC) bits during the last decade indicated the possibilities for serious improvements in drilling performance, but they also required higher rotational speed along with high torque. These parameters can be provided at the same time by using the new generation of DHMs (see Chapter 2).

Development and implementation of electrodrilling technology for drilling single and cluster vertical and directional wells

In 1941, Russian engineers and designers built the first efficiently working electrodrill that was used for drilling wells with TDs of 1500 m. The successful application of turbodrills in Krasnokamsk stimulated work toward further improvement of electrodrill technology, because of its potential merits compared to the hydraulic motor. Some of these qualities are presented as follows.

- the constant bit weight while drilling regardless of the parameters of drilling mud circulation
- the possibility of monitoring bit performance and condition from the surface, as well as controlling the rock destruction process
- the possibility of obtaining information about the properties of the drilled rock by analyzing of the bit performance

However, achievement of these advantages hinged upon the successful solution to certain technical problems related to channeling the two parallel energy streams to the bottomhole: the electrical current for the bit rotation and the fluid or air flows for cleaning the bottomhole.

As a result of scientific research and engineering work that was resumed in 1947, the designs of the electrodrill, the power cable, and other auxiliary tools and instruments were significantly improved. From 1948 through 1951, the first pilot commercial operations using electrodrills took place in Azerbaijan and Bashkiriya. As the whole electrodrilling system was improved and the normal series of motors was developed from 1952 to 1963, electrodrilling became widely used in oilfields of the Ukraine, Kuibyshev (now Samara) region, and Turkmeniya.

The experience gained from the extensive application of electrodrilling during the decade (nearly 500,000 m) in various geological conditions allowed the evaluation of the merits and shortcomings of electrodrill technology. The results of drilling several wells in Bashkiriya using electrodrills were 15–20% higher compared to the results of turbodrilling. However, lack of steady improvement of these results failed to stimulate the wide use of electrodrilling, which would require setting up special service and repair shops, as well as organizing advanced professional training of electrical engineers. Certain additional work was required to make substantial improvement in the electrodrilling system and, most important, to identify the most efficient areas of application where electrodrills would have a significant advantage compared to HDHM.

This work was performed from 1963 through 1970 by VNIIBT in cooperation with and incorporating significant input from the Special Design and Technological Bureau of Submersible Electro-motors, including Electrodrills (SKTBE) of the Ministry of Electrical Engineering in Kharkov, Ukraine. The work resulted in the development of improved design 127–290-mm outside diameter (OD) electrodrills, with speed-reducers capable of reducing rotational speeds to 70 rpm, and a borehole deflecting tool. The primary achievement of this work was the development and introduction of the world's first successfully operating cable telemetry system (STE). The system used the power cable to transmit signals to the surface, which allowed receipt of real-time information about the borehole path parameters, such as the direction and inclination, as well as the deflector position. Operation of the new system promoted the expansion of the application of electrodrills for drilling directional and horizontal wells.

Electrodrills successfully drilled wells to depths in excess of 5000 m with complicated geological conditions that necessitated the use of heavy mud. Utilization of electrodrills when drilling with foam, air, gas, or aerated mud also indicated that, in many cases, they were unrivaled. Detailed information about electrodrilling is presented in Chapter 2 and in the second volume of this series.

Improvements in electrodrill technology resulted in increased volumes up to 600,000–1,000,000 m per year (*see* the diagram in Figure 1–8b). When correctly applied, electrodrilling applications can recoup the additional operating costs the method requires.

The comparatively low proportion of wells drilled using electrodrills can be explained not only by the need for qualified personnel and special facilities to service the equipment, but for purely political reasons. Certain high-ranking officials from the U.S.S.R. Oil Ministry used their influence to block the development and implementation of this method. However, development of electrodrilling technology continued, thanks to the efforts of advocates of the method such as N. K. Baibakov, the FSU State Planning Committee of the U.S.S.R. (GOSPLAN) Chief, the specialists from the VNIIBT and the SKTBE Institutes, and some oil companies, the best example of which was Bashneft.

The development of the oil and gas well drilling technology in the FSU during the postwar period featured application of the three drilling methods that are summarized in sections of this chapter.

- rotary drilling
- turbodrilling
- electrodrilling

The predominance of downhole mud motor technology has radically influenced the development of the oil-producing industry in the FSU. Development, improvement, and implementation of the new trends of drilling technology involve the wide application of downhole mud motors. Use of the DHM technology encouraged a large number of new trends in the development of drilling methods. Some of them seem to preserve the interest of the oil industry specialist. The main trends are outlined as follows, and chapters in Volume 2 provide detailed descriptions.

New trends in drilling technology and commercial application of DHMs

Horizontal and multilateral borehole drilling. In 1953 in the Ishimbai region of Bashkiriya, a multilateral well was drilled with 10 horizontal boreholes. In 1973, the electrodrill method was used to drill a multilateral well with five horizontal boreholes in the Dolina oilfield in the Western Ukraine. Later, this trend was used widely in many regions of the FSU and in the West. Electrodrill technology was the most promising because it enabled the use of gaseous agents and foam while drilling through the productive zone.

Extended reach drilling (ERD). In 1968, an exploratory well was drilled near the Markovo village in Eastern Siberia. The actual well depth was 2800 m, whereas the true vertical depth was 2100 m, and the length of the horizontal section was 632 m. An additional directional hole was sidetracked from the main horizontal interval to penetrate through the productive zone.

Horizontal boreholes sidetracking through a cased hole. In 1969 in the Zybza field in the Krasnodar region, a directional hole was sidetracked through a casing window in well No. 90. The directional borehole with a maximum inclination of 81°, and a length of 290 m was drilled through the productive zone.

Volume 2 in this set presents comprehensive data on directional and horizontal drilling technologies.

Aluminum drillpipe (ADP). The first research work and experiments with ADP, started in the second half of 1950s, produced positive results. During the following years, ADP was widely used, especially in drilling directional wells combined with both hydraulic and electrical motors. ADP accounted for 60% of the total drillpipe consumption. Wide use of ADP enabled a substantial reduction of loads applied to hoisting equipment on drilling rigs, reduced trip time, lowered hydraulic losses in the borehole, and reduced mud pump loads, as well as lower transportation costs. ADP was successfully used in rotary drilling as well. (See Chapter 2 in the second volume for more ADP details.)

In addition, aluminum alloy was used to manufacture casing pipe. Aluminum casing string has been successfully operating in one well for nearly 20 years.

Tubing-conveyed, small HDHMs for well workover and completion operations. The first successful pilot-commercial work was done in 1940 and 1941 in Baku. The work involved running a 125-mm turbodrill, 140-mm two-cone bits, and some other tools through the 6-in. production string. After their introduction, PDMs were used for the same purpose in smaller size casing strings. This technology proved especially efficient in offshore producing wells where the use of tubing combined with DHMs rather than drillpipe and workover units significantly improved the economics. [13]

Drilling 394–1000 mm diameter upper intervals of deep wells using the rotary-turbodrill system (RTB). The RTB is essentially a monolith unit connecting twin-turbines (parallel assembled). The RTB technology has many advantages compared to rotary drilling using large-size drillcollars. The large diameter of the unit, along with high rotational speed and eccentricity of drillbits, provides a high penetration rate at a lower bit weight, which enables drilling nearly vertical boreholes. This method of drilling improves borehole conditions and, consequently, the quality of casing and cementing jobs.

In 1967, RTBs were used in the oil and gas production industry for the first time (earlier they were used by the mining industry) to drill well SG-2 Biikzhal in Western Kazakhstan with a TD of 7000 m. The RTB was successfully used to drill two intervals of the well. RTB8-920 drilled the first interval with a diameter of 920 mm from 18–362 m. The interval with the zero inclination was successfully cased with a 720-mm welded casing string and cemented. RTB2-640 drilled the second 640-mm diameter interval of 362–1369 m. The interval with an inclination of 1°30' was cased by a 426-mm string. This successful experience gave strong impetus to the use of RTBs for drilling the upper section in hundreds of deep wells in various regions throughout the FSU, from the Western Ukraine to the republics of Middle Asia, including several super-deep scientific boreholes. RTB technology details are presented in Volume 2.

Drilling without pulling out pipe with HDHMs and retractable drillbits (RBs) casing drilling. From 1948 through 1952, several experimental intervals in some wells were drilled near Krasnokamsk and in the Saratov region using the casing drilling technique. The method involved use of the 8⅝-in. (219 mm) casing string and a pilot expandable underreamer (EUR). The diameter of the interval was 295 mm. About 1000 m were drilled in three wells, proving the operating capability of the drilling method.

During the following years, engineers designed, fabricated, and tested various modifications of the retractable two-cone and tri-cone bits. From 1964 through 1975, the Saratovneftegaz Company used these bits in the lower intervals of wells with TDs of 2000, 2500, and 3000 m. The new method was used to drill 30 wells with a total penetration of 50,000 m. Wells with diameters of 295 mm and 220 mm were drilled using 8⅝-in. and 6⅝-in. (168 mm) special pipe. Meanwhile, new drilling technology and equipment was developed and used. Among these were special thin-walled tool joints for casing pipe with stabilizing shoulders, a retrievable tools system for running in and pulling out of the hole using straight and reverse circulation of drilling mud.

In the mid-1970s, Western and Russian geoscientists introduced a new method to study Earth's crust. The method was to drill ultra-deep wells with continuous coring, which required a large number of trips for core recovery. From 1975 to 1979, Russian engineers and designers developed a system of retractable tools for drilling in crystalline rock. A new drilling technique that enabled drilling without pulling out a drillstring was tested in a satellite borehole near the Kola ultra-deep hole. From 1985 through 1992, the method was successfully used to drill a section in the Krivoy Rog super-deep well to a depth of 5450 m with continuous coring.

Since 1991, the method has been used to drill offshore scientific and stratigraphic boreholes (detailed description in the last chapter of Volume 2).

Super-deep well drilling for scientific purposes. Drilling was completed in 1983 on the world's deepest SG-3 borehole on the Kola Peninsula. The hole was drilled to a depth of 12,242 m.

The borehole penetrated through the basement crystalline rock to obtain information on the structure and composition of the deep zones of the ancient continental Earth crust. The borehole was drilled with continuous coring using various multi-section and reduction gear turbodrills as well as ADP. A similar technology is currently used for drilling scientific wells in crystalline rock (*see* Volume 2).

Main Components and Management of Drilling Operations, Personnel Training, Scientific Research, and Design Work in the FSU

Exploratory and key stratigraphic drilling

The centralized management structure, set up after nationalization of the oil industry, contributed greatly to the successful development of the industry. Besides significant progress in the development of drilling technology and equipment, the system of drilling operations management also contributed to its success. First, the geological service was set up. This was an extremely important step that enabled the systematic growth of exploratory drilling volumes and the development of new methods for exploration activity. In addition to the need to expand the amount of discovered and recoverable oil reserves, the need for such development was dictated by the requirement to work out certain methods that would resolve the problem of production from flooded oilfields. These were primarily the fields near Baku and Grozny that had been flooded in the early 20th century.

Under the Russian government, the Geological Committee received more authority. The committee had existed before the revolution of 1917 but had little influence in the regions. A special department for hydrocarbon exploration was established within the committee. The relevant geological services were organized in the oil-producing regions. Drilling companies boosted exploration drilling and increased its proportion in total oil well drilling. In 1923 and 1924, only 360 m of exploratory wells were drilled, whereas in 1927 and 1928, 69,300 m were drilled. Exploratory drilling increased correspondingly from 0.3% to 19.1%.

This tendency increased even more in the following years. The total meterage of drilled exploratory wells was 124,400 m in 1929, and by 1932, this had grown to 239,500 m. The corresponding exploratory percentages were 27.9% and 32.2%. Still, exploratory work was conducted primarily in the old fields, which involved drilling and completing in deeper horizons. The proportion of exploration activity in new fields was relatively low.

In the following years, more exploration work was done in new areas, especially in the Urals-Volga oil and gas province. Total meterage of exploratory wells in 1937 was 460,700 m, and in 1940, it amounted to 501,800 m. This did not slow down with the discovery of a large number of new fields. Between 1920 and 1933, 15 new fields were discovered in the FSU, whereas 12 fields were discovered in 1938 alone. From 1934 to 1939, 47 new fields were discovered, including many large oilfields with high potential. Among them was the Kara-Dag field near Baku, the Starogroznenskoye field near Grozny, the Tuimazinsky field in Bashkiriya, the Syzranskoye field in the Samara region, the Krasnokamskoye field in the Perm region, the West Nebit-Dag field in Turkmeniya, and the Andizhan field in Uzbekistan. The postwar period witnessed even more intensive exploration drilling (Table 1–2) [14]

TABLE 1–2
Main Characteristics of Exploratory Works After WWII (1940–1955)

Year	Drilled, Thousand Meters			Areas in Exploration		Drill Rigs	
	Total	New Areas	% of Exploratory Drilling	Total	New	Total	Exploratory
1940	501.8	137.5	27.3	133	55	399	172
1945	383.0	183.5	42.6	139	97	262	160
1950	1,980.0	1,280.0	49.7	379	376	1,079	800

In addition to the significant growth of exploration drilling, the oil industry in the FSU marked other achievements. Among them was the program of drilling deep key stratigraphic wells that was implemented in the first decade after the war. The program was initiated and developed by the academician I. M. Gubkin, the founder of the highly developed geological service in the FSU. The wells were drilled to the crystalline basement, and the objectives were to obtain information about the deep subsurface structure of the country's subsoil and to develop oil- and gas-prospecting programs. As a result of key stratigraphic drilling, Russian petroleum geologists received an enormous amount of data about the FSU subsurface geology. This data allowed a better evaluation of oil and gas prospecting work in various regions of the vast country. Table 1–3 gives an idea of key stratigraphic drilling activity. [15]

TABLE 1–3
Key Stratigraphic Wells Drilling
After WWII (1946–1955), Thousand Meters

	1946–1950		1951–1955	
Areas	**Drilled Thousand Meters**	**Completed Wells**	**Drilled Thousand Meters**	**Completed Wells**
Caucuses	32.0	5	47.1	24
Urals	24.2	3	13.3	12
Central	43.4	25	44.8	28
Siberia	15.0	1	71.8	26
Other Regions	31.0	6	42.4	22
Total	145.6	40	219.4	112

The information obtained from key stratigraphic drilling brought new explanations about the deep subsurface structures in many regions and clarified the structure of both the Russian platform and its entire depositional sequence. Additionally, geologists received data from stratigraphic drilling in the southern part of the Eastern Siberian Plain. In some regions, stratigraphic drilling identified new structures as potential targets for deep exploratory drilling. Besides these achievements, this period witnessed a significant growth of structure test drilling with coring that played an important role in preparing the structures for exploratory drilling. In 1946, 214,200 m of cored wells were drilled, and in 1955, cored wells totaled 2,050,000 m. About 800 structures were identified during this period as targets for further exploratory drilling.

The rapid development of exploration activity in the postwar period from 1946 through 1955 resulted in the discovery of 254 new oil and gas deposits in previously developed fields. The amount of proven recoverable reserves increased fivefold during this period. The newly discovered oil areas included such major fields as the Bavlinskoye and the Romashkinskoye fields in Tatariya; the Mukhanovskoye in the Samara region; the Shkapovskoye and the Belebeyevskoye fields in Bashkiriya; and the Tashkalinskoye and the Ozek-Suatskoye field in the

Grozny region. The Berezovskoye field—the first large gas field—was discovered in Western Siberia. Further increase of exploratory drilling volumes in Western Siberia in the second half of the 1950s and early 1960s led to the discovery of such major oilfields as the Ust-Balykskoye, the Surgutskoye, the Nefteyuganskoye fields, and the giant Samotlor field, one of the world's largest.

These discoveries show that the visionary policy of the FSU government, the management of the Oil and Gas Ministry, and the Ministry of Natural Resources in the postwar period helped build the large proven reserve base. The discoveries have stimulated rapid development of the industry and still contribute much to the successful work of oil industries in Russia and other Commonwealth Independent States (CIS) countries despite the substantial decrease in the amount of exploration drilling in subsequent years.

After World War II, the government introduced a vertical management structure in the FSU Geological Survey and streamlined the organization's work. The government formed the U.S.S.R. Ministry of Geology that governed the work of the Ministries of Geology of the FSU Republics. The Ministries of the Republics controlled the exploration activity in mineral ore and hydrocarbon products. The structure of the Ministries of Geology included exploration drilling companies, geological and geophysical services, and organizations that participated in prospecting for oil and gas. The exploration drilling companies formed geological exploration expeditions and individual exploration groups.

The major task of the country's and the republics' Ministries of Geology was to discover and explore oil and gas fields in remote areas that had not been covered by oil and gas producing companies. For example, the companies and organizations of the Russian Federation Ministry of Geology carried out initial exploration work and discovered a number of fields in Siberia. The oil- and gas-producing companies, set up in the explored regions, worked on additional appraisals of the identified structures and completion of the deeper productive horizons. The producing companies had their own geological exploration services.

Successful development of geological exploration activity in the FSU and the breakthroughs achieved in this field owe much to a group of people who used their talent and effort to enable these achievements. This group included prominent geologists and mining engineers with experience from the pre- and post-revolution period. Among them were I. M. Gubkin, S. M. Androsov, M. V. Abramovich, D. V. Golubyatnikov, A. Ya. Krems, A. A. Bakirov, N. B. Vassoyevich, K. A. Mashkovich,

M. B. Mirchink, A. N. Mustafinov, V. S. Melik-Pashayev, P. A. Trofimuk, N. S. Yerofeyev, K.Y. Ervye, E. A. Kozlovsky, and F. K. Salmanov.

Management of drilling operations in the oil and gas industry

Similar to the geological exploration sector, the oil-producing industry also had a vertical management structure. The Ministry of Oil Industry included Main Directorates of Oil Production that controlled production work in certain territories. These directorates included Deputy Drilling Managers and drilling departments consisting of qualified and experienced specialists. The Chief Engineer and the oil production department specialists managed routine issues of the oil production process. The Main Directorates included the Equipment and Materials Procurement Departments, the Planning and Financial Department, and the Accounting Department.

The oil-producing associations, formed in the oil-producing regions, reported directly to the Ministry's Directorates that also included drilling and geological services. The oil-producing associations included producing trusts that, in turn, consisted of a drilling bureau. The associations that carried out a large amount of exploration drilling included oil and gas exploratory drilling trusts with their own drilling exploration bureau.

The Ministry of Oil Industry included the General Geological Department as well. The task of this unit was to coordinate and inspect the regional geological services and to define the trends of exploratory and development drilling patterns in the explored fields.

One distinctive feature of field development stands out against the background of the general dissimilarity between management of drilling operations in the FSU and in the West. This is related to the centralized state management and planning of the industry's activity in the FSU. Only one drilling contractor carried out drilling activity on a discovered field during the entire period of field development after its delineation, whereas in the West several independent contractors may be drilling wells in the same field. This situation, from our viewpoint, may reflect negatively on the general drilling strategy and field development on the whole.

The FSU government and officials in the Ministry of Oil Industry understood the essential role of drilling in the oil-producing industry and paid strict attention to its development. The issue of drilling received high priority among other activities in the oil industry. Significant capital investments were made in drilling operations (see Table 1–4). [16]

TABLE 1–4
FSU Capital Spending in Oil Industry and Well Drilling (1923–1955)

Period	Total Capital Spending in Oil Industry (Mil Rubles)	Including Well Drilling (Mil Rubles)	% of Total Volume
1923–1927	1,047	450.4	43.0
1928–1932	2,151	685.7	31.9
1933–1937	4,304	1,586.0	36.4
1938–1940	3,487	1,011.0	29.0
1941–1945	5,566	1,008.6	18.1
1946–1950	22,898	7,681.1	33.5
1951–1955	53,066	18,720.8	35.3

The dynamics of capital investments provided the impetus for successful exploration and development drilling. In 1922, 4.7 million tons of oil were produced, whereas in 1941, production was 33.0 million tons; in 1958, it was 38 million tons; in 1960, it was 160 million tons. The pace of oil development in Western Siberia resulted in the peak oil production in the FSU—624 million tons in 1988 (*see* Fig. 1–6).

Occupational training in the oil industry

The rapid development of the oil and gas industry after the October Revolution of 1917 and after World War II resulted in increased demand for highly qualified specialists for the industry. The issue was resolved by a number of factors. First, most of the mining and oil industry specialists stayed in Russia after the revolution. They were able to share their experience with the young engineers and technicians during the post-revolution period.

Second, the government placed an emphasis on improvement of the general educational level of the population by fighting illiteracy. Before 1917, the rate of illiteracy in Russia was 70%. However, by the 1930s, the situation had drastically improved, and in the 1940s, practically the entire population in Russia was literate. The number of educational establishments significantly increased, such as secondary schools (especially in the rural areas), vocational training schools, and colleges and universities. The universities usually had part-time education programs for working people as well as correspondence training programs for students who lived in cities and villages that did not have universities or colleges. Secondary and higher education were free, thus providing a major incentive for students to improve their level of competence.

A large number of specialized mining and petroleum schools and universities were formed in the country. Among the oldest and largest universities were the Azerbaijanian Industrial Institute in Baku, the Petroleum Institute in Grozny, the Moscow Petroleum Institute named after Gubkin, the Moscow Institute of Geology and Exploration that was spun off from the Moscow Mining Academy, and the Mining Institutes in Leningrad (St. Petersburg) and Sverdlovsk (Ekaterinburg).

Some major universities, such as Moscow State, St. Petersburg, and Ekaterinburg had mining colleges. All these measures resulted in a larger number of young professionals who graduated in the late 1920s and early 1930s. Later, some of them became prominent managers and founders of drilling companies and outstanding scientists and academicians in the oil and gas industry.

Scientific work

Along with the formation and development of drilling companies in the postwar period (in the second half of the 1940s through the 1950s), a large network of scientific and research institutes and design bureaus were set up. These organizations undertook the development of new equipment and technology for drilling oil and gas wells. In 1949, three special design bureaus (SKB) were formed: SKB-1 for the development of electrodrills, SKB-2 for the elaboration of new turbodrill and bit designs for high rotation speed drilling, and SKB-3 for the development of the cable electrodrilling system. The VNIIneft Scientific and Research Institute had laboratories that developed the composition of drilling fluids and cement slurries. The Institute of Fossil Fuels (the former Institute of Oil) also had laboratories that concentrated on the development of drilling process stages.

In 1953, all the scientific, research, and design bureaus and organizations located in Moscow were joined together to form the world's first All-Union Scientific and Research Institute of Drilling Technology, VNIIBT. The institute's activity included the development of all issues related to the oil and gas well drilling process. The Institute intensively developed and implemented advanced drilling equipment and technologies. The VNIIBT had a subsidiary in Perm and more than 10 groups working in other oil-producing regions in the FSU. The Institute's assets included experimental plants in the Moscow and Volgograd regions and a large test facility in the Moscow region. The test base was equipped with full-size rigs for drilling test wells in blocks of rock (mainly marble and granite). The 500-m depth cased borehole was also available for experiments there.

At present, the Institute has been transformed into the joint-stock company (JSC), NPO Burovaya Tekhnika, which preserved the VNIIBT structure, with the exception of groups outside Russia. The main activity of the new company includes the development and fabrication of drilling tools, such as drillbits (including PDC type), core barrels, packers, PDMs, and others. The company also provides directional drilling services including well design and certification testing of drilling tools and equipment. [17]

In addition to the VNIIBT, the Regional Scientific and Research (NIPI) were formed in all major oil and gas producing regions. The institutes undertook well planning and drilling research with special attention to the regional specifics.

For many years, VNIIBT coordinated the activity of these institutes related to drilling issues. In addition to the Research Institutes, several large engineering and design institutes for the development of drilling and production equipment were set up. Among them were institutes such as Giproneftemash in Moscow and AzINMASh in Baku that provided engineering plants with the necessary documentation for the new equipment prototypes. These plants included a group of equipment manufacturers, such as the plant named after Schmidt, the Uralmash in Ekaterinburg, and the Barrikady plant in Volgograd. These engineering plants had their own large design bureaus that developed and designed different drilling equipment.

Professional drilling engineers appeared in the oil patch in the post–World War II period. Their knowledge base served as a source of original ideas as well as scientific and research developments for the creation of a number of advanced drilling technologies. Unfortunately, some of these ideas were not used by the drilling industries in the FSU and abroad for reasons that are highlighted later in this chapter. However, information about these ideas and technical solutions will be of interest and use, since many of them are relevant even today.

Despite the achievements in the FSU oil industry, certain negative trends appeared in the course of its development. As time went by, these negative trends became stronger and had a negative effect on the development of the industry. The initial positive role of centralized planning and management of the oil industry turned negative. The most significant disadvantages are described in the following section of this chapter.

Negative Trends in FSU Drilling Industry Developments

Significant decline in the growth of geological exploration

After the rapid development of the oil industry that started in 1960, exploration activity declined, which resulted in a substantial decrease in the proportion of exploratory oil and gas wells compared to the total number of wells drilled.

The proportion of exploratory drilling from 1955 to 1990 is shown here:

1955	43%
1960	51.9%
1970	43%
1975	35%
1980	26.4%
1985	19%
1990	16.9%

In the following years, this proportion declined further. Drilling rates of exploratory wells were significantly lower than the rates of development drilling. This was most regrettable for the Western Siberia oil and gas province, which is, and will remain in the foreseeable future, the largest oil-producing region in Russia.

The proportion of exploratory drilling in 1990 that was carried out by companies controlled by the Ministry of Oil Industry was as low as 1.15%. Additional exploration drilling by companies reporting to the Ministry of Geology resulted in a slight increase to 4–5%. Such a shortsighted policy led to a decline in the amount of proven recoverable reserves. At the same time, the ratio between annual oil production and the amount of reserves became unacceptably low compared to the normal 20–25 year average. According to information from *World Oil* magazine, [18] recoverable oil reserves in the FSU went down during the period between 1982 and 1990 from 13.7 billion cubic meters to 10.05 billion cubic meters (27%).

At the same time, the proportion between production and reserves was equivalent to about a 13-year average. During the next few years, this proportion continued to decline until it stabilized due to a decline in production rather than growth in hydrocarbon reserves. This could be explained by insufficient investment in the oil industry as well as euphoria on the part of top industry and government officials caused by previous successes in the industry.

Insufficient capital investments

Plans called for a large increase in oil production, which required related growth in development drilling volumes, but the limited amount of investment curtailed exploratory drilling. Although revenues from oil exports amounted to billions of petrodollars and were the main source of hard currency earned for the country's budget, the industry received a very small share of those revenues. This portion was insufficient to import needed equipment and materials that were unavailable domestically.

The main reason for this state of affairs was the priority given to financing the military industry for several decades, which consumed large amounts of money during the Cold War. Another large portion of the country's budget went to support the governments in countries building socialism, such as Cuba, Vietnam, and Eastern European countries. Significant resources were used to finance Communist parties in a number of countries as well as the military action in Afghanistan.

The decline of capital investments also affected the civil engineering and metallurgical industries. As a result, the fixed assets of some of the engineering plants that manufactured oilfield equipment were not replaced for several decades. This negatively affected the quality and quantity of their production and resulted in lagging development and production of advanced equipment. During these years, drilling companies in the FSU incurred many losses due to downtime as well as the shortage of imported high-strength casing and drillpipe with new pressure-tight thread connections, durable tri-cone bits, BOP equipment, chemicals, and other equipment.

The quality of drilling rigs made by the Uralmash and Barrikady plants was significantly lower compared to the quality of American-made rigs. These factors included the load capacity, maximum allowable mud pump pressure, efficiency of the mud-cleaning system, and durability of rig parts and assemblies. Some diesel-driven rigs used aircraft and tank engines, since the domestic producers did not make low-speed, high-capacity engines with the long service life that was required by the Russian drilling industry. This complicated the design of drilling rigs used for exploratory drilling in areas without an available power supply. It made them heavier and raised drilling costs.

These circumstances contributed greatly to a substantial level of downtime for rigs and crews and accounted for 18–20% of the total drilling time.

In total, during the period from 1965 to 1990, the level of annual drilled footage increased 9.16 times. Of this increase, 50% was due to the fact that the number of operating drilling rigs grew from 782 to 2353. The remaining 50% increase in drilled footage was due to a higher average overall drilling rate (in Russia it's called *commercial rate of drilling* or CDR) that went up from 495.5 m/rig-month to 1508.7 m/rig-month. In other words, the development of the industry had a semi-extensive and semi-intensive nature. However, this conclusion is only true on the surface.

The principal increase in overall drilling rate was achieved because of the significant growth of drilling volumes in Western Siberia. The region features favorable geological drilling conditions: the presence of large intervals composed of easily drillable sandy-argillaceous deposits, the absence of horizons with abnormally high pressure, no lost circulation problems, and an average well depth between 2500 m and 2700 m. These factors contributed to the CDR increase from 1500 in 1965 to 5000 m/rig-months in 1990 (3.3 times growth). Drilled meterage grew from 190,000 m to 25,000,000 m (132.6 times growth).

In other regions of the FSU during this 35-year period from 1955 to 1990, the average CDR increased from 495.5 m/rig-month to 784.7 m/rig-month, or 1.58 times. During the same period, the number of meters drilled annually went from 4,650,000 to 17,400,000, or 3.74 times, whereas the number of operating rigs grew at a rate of 2.36 times.

The differences in drilling results achieved in Western Siberia and in the older oil-producing regions of the FSU can be partially explained. The major portion, 7,000,000 m out of 7,200,000 m, of the exploratory drilling in 1990 took place in the old regions. Most of the deep and ultra-deep wells that normally accounted for rather low CDR (250–300) were drilled in these regions.

Along with a lack of financial investment, other factors contributed to the negative trends in the development of the drilling industry, such as an inefficient production pattern and low labor efficiency under the existing economic system.

Rigid planning of drilling activity

In the rigid standardized state plans, one of the main indicators of drilling activity was the amount of drilled meters.

Rigid planning was unlike the normal planning used in most countries. Instead of being based on scientifically grounded forecasts and recommendations, planning in the FSU was made in the form of law and order. The government's plan of industry development was adopted as a law. Failure to comply with and implement this law could sometimes entail extremely serious consequences for the company's top management. This dogmatic approach to planning completely reversed the essence of sound action. This was especially true for planning that set forth faulty objectives, as in the case of drilling activity.

In addition, the levels of wages and bonuses fully depended upon successful implementation of the plan. As a result of this "race" for meters, drilling companies used to start drilling before the rig installation was completed or the drilling equipment fully tested, which caused failures and downtime and affected the quality of the well.

While drilling an open hole through high pressure horizons, the mud properties did not correspond to what was required due to lack of weighting agents, such as barite, hematite, and other chemicals. In addition, often this drilling was done without having prepared and tested casing pipe available at the rig because of the shortage of casing pipe. Frequently, the casing string was run too late, when the open hole encountered problems, which did not allow the casing shoe to reach the planned depth. The quality of cementing jobs and operations to isolate water- and gas-bearing horizons was poor. All these factors led to uncontrolled oil and gas shows and blowouts.

The requirement to report the results of the plan fulfillment to the government prompted drilling companies to falsify information reported by overstating the meterage actually drilled. Sometimes a specialist from a drilling company even bribed the geophysical contractor specialists to have them falsify well directional surveys to report that the well had reached the target. This caused a decrease in field production, premature water influx, and other negative events.

In addition, this type of planning significantly complicated the work on testing and implementation of new well drilling technologies and equipment. Exposed to the pressure from existing drilling plans, managers of the drilling companies and toolpushers opposed the program of new equipment testing, since they had no confidence in the positive results of the experiments. In many cases, they would start the test under orders from top management but discontinue it at the earliest opportunity.

This was the case with testing of various drilling equipment, such as turbodrills, directional drilling tools, electrodrills, and some other innovative drilling tools that later became widely used by the industry. There were many examples when the oil-producing companies and drilling organizations failed to find an application for promising and efficient technologies and methods that were developed, designed, and fully tested with very positive results.

A good example of the companies' reluctance to apply new technologies was their unwillingness to use horizontal well drilling which is currently one of the most popular techniques used around the world.

As previously mentioned, the world's first multilateral horizontal well was successfully drilled in 1953 by a drilling group from the Ishimbaineft Company from Bashkiriya, headed by A. M. Grigoryan, the drilling engineer. After that, in 1957 and in 1960, more multilateral horizontal wells were successfully drilled in Borislav (West Ukraine) and Krasnodar. In 1968, the same group drilled a horizontal well near Markovo, a village in Eastern Siberia, which set the world record for horizontal section length—632 m.

Despite the proven efficiency of this method for completing the productive zone, both the drilling companies and Ministry officials were reluctant to promote its wide application and further development. The Ministry management was concerned that an increased investment would cause GOSPLAN (the state planning agency) to curtail budgets and impose higher oil production levels if horizontal drilling was widely used by the industry. At the same time, the oil-producing bureau management was afraid to see the same actions from the oil companies. They all failed to understand that horizontal drilling requires much attention and involves an additional amount of work to drill and complete a well.

Hence, for more than 20 years, the new method of oilfield development that could significantly increase well flow rates and enable more oil recovery was not used in the FSU where it was developed. Only in the late 1980s did the Russian drilling industry follow the example of the United States and other countries and begin using this technology in various regions (see Volume 2).

Unfortunately, the list of such examples is quite long.

Absence of private property ownership, private entrepreneurship, and initiative

One of the fundamental disadvantages of state ownership of all means of production and productive forces was that a producer did not own the products he made. Whether he was very efficient or totally inefficient had little impact on his wealth, which hinged only on his salary level as prescribed in the payroll table and a small bonus if production targets were achieved. Therefore, any improvements in production processes due to new technology did not affect his financial position.

At the same time, testing and implementing new equipment and technology required extra effort, initiative, and sometimes involved risk. The absence of compensation for these efforts substantially slowed the development of drilling technology. Of course, the industry saw many hardworking professionals with marked initiative who tried to implement new technology. Yet, the enthusiasm shown by some individuals did not always lead to acceptance or use of the innovations throughout the industry.

A typical example of this situation was the development and application of electrodrilling technology for drilling multilateral horizontal wells in the Dolina oilfield in Western Ukraine. Volume 2 describes this work in detail. The 11 multilateral horizontal wells were drilled under the technical guidance of VNIIBT engineers. Most of the wells were successful, however, the management of the company that developed the field kept well production information carefully under wraps and was very unwilling to release the information even to the VNIIBT specialists who designed the wells and managed their construction. The institute was unable to obtain any data for the most successful well, No. 350. Only the production foreman shared some information about the actual flow rate on the condition that this data would not be released.

During an informal meeting with the company's management, one of the authors of this book (Y. A. Gelfgat) asked questions about the reasons for such secrecy and the unwillingness of the company to drill more of these wells. The manager of the company answered that, if he had informed top management about such high flow rates, the oil production level planned for the following month would be increased, whereas he was not sure whether new wells would be as successful. He also cited the lack of incentive for him and his people to put in the additional effort to complete these wells and raise oil production.

In addition, he said that since the work on commercial implementation of the new method would take several years and he was going to retire soon, he did not see any personal benefit for promoting such work. At the same time, he described an imaginary situation in which he was the owner of the company that developed this field and could hand it down to his son. In this case, his motivation would be sufficient to advocate this work strongly. These reasons were hard to argue with. Even advocates of a socialist society and the elimination of private ownership would have had a hard time coming up with sound counter arguments. Consequently, the work on drilling multilateral horizontal wells at this field was halted.

Did a new era begin in Russia?

In 1985, perestroika (or reorganization) was started in the FSU, which, up to 1991, was mainly related to the government's domestic and foreign political activity. The following years witnessed economic reforms such as privatization of industry, introduction of a free market economy, and elimination of government price controls.

However, economic reforms in the first several years of the transition period failed to improve the Russian economy. According to information from the Russian Federation Committee on Statistics, the level of oil and gas condensate production dropped to 305 million tons in 1997, gas production fell to 544 billion cubic meters, and only 8,300,000 m were drilled.

It is difficult to predict when the situation will change, but the transition to private ownership of production will eventually yield positive results. During the transition from the FSU to the Russian Federation, vertical integrated oil companies were developed: Lukoil, UKOS, Surgutneftegaz, Sibneft, TNK, Rosneft, Slavneft, and Sidanco, and regional oil companies were strengthened Tatneft, Bashneft, and others.

Since 1996, production rates increased (see Table 1–5), but drilling activity had a tendency to change. It was decreasing in 1998, but since 1999, drilling activity started growing again (see Table 1–7). Several publications provided a good opportunity to follow the story of modern developments in the oil and gas industry. *World Oil* magazine, regular publications of the FSU/Eastern Europe Reports, and other sources (primarily *Oil & Gas Journal*) have published articles on Russian drilling activities/technologies along with new Russian publications like *Neft I Kapital (Oil & Capital)* magazine. Tables 1–5, 1–6, and 1–7 summarize some of the data on production and drilling activity in Russia during the last decade or more.[19] [20] [21] [22] [23]

Evidence of structural changes in the drilling industry is clear, but the formation of independent drilling contractors and integrated service firms still faces difficulties. The primary goals of the new oil companies were to enhance production with minimal investments, but the time has come to renew investments in drilling and exploratory activities. We strongly believe that research capital gained by the Russian drilling industry in the past should be unconditionally demanded for this purpose.

TABLE 1–5
Crude Oil and Condensate Production in FSU Countries (1994–2000)

	Barrels per Day						
Country	**1994**	**1995**	**1996**	**1997**	**1998**	**1999**	**2000**
Russia	6,280,100	6,090,000	5,980,000	5,907,602	6,041,071	6,070,948	6,451,000
Kazakhstan	434,000	409,186	462,817	518,007	571,232	604,172	700,320
Remainder of FSU	502,369	541,350	564,278	577,18	586,297	600,947	687,246
Armenia	0	0	0	0	0	0	0
Azerbaijan	192,100	182,473	190,000	179,698	180,291	179,105	278,298
Belarus	40,234	38,481	36,948	36,269	36,466	36,649	36,600
Estonia	0	0	0	0	0	0	0
Georgia	1,800	1,500	3,000	1,992	2,190	2,180	
Kyrgizia	1,640	2,000	1,990	1,694	1,535	1,530	2,179
Latvia	0	0	0	0	0	0	0
Lithuania	1,655	3,097	3,235	4,231	5,474	4,641	6,303
Moldova	0	0	0	0	0	0	0
Tajikistan	1,000	2,000	1,800	520	400	395	366
Turkmeniya	75,700	78,500	86,805	111,130	121,250	139,425	141,030
Ukraine	79,240	79,299	69,000	84,560	77,679	75,688	73,495
Uzbekistan	109,000	154,000	171,500	157,093	161,012	161,334	148,975
Total FSU	7,216,469	7,040,536	7,007,095	7,002,796	7,198,600	7,276,067	7,838,566

	Thousand tons per year (1 bpd = 49.8 ton per year)						
Country	**1994**	**1995**	**1996**	**1997**	**1998**	**1999**	**2000**
Russia	312,749	303,282	297,804	294,199	300,845	302,333	321,260
Kazakhstan	21,613	20,377	23,048	25,797	28,447	30,088	34,876
Remainder of FSU	25,018	26,959	28,101	28,744	29,198	29,927	34,225
Armenia	0	0	0	0	0	0	0
Azerbaijan	9,567	9,087	9,42	8,949	8,98	8,919	13,859
Belarus	2,004	1,916	1,840	1,806	1,816	1,825	1,823
Estonia	0	0	0	0	0	0	0
Georgia	90	75	149	99	109	109	0
Kyrgizia	82	100	99	84	7	76	109
Latvia	0	0	0	0	0	0	0
Lithuania	82	154	161	211	273	231	314
Moldova	0	0	0	0	0	0	0
Tajikistan	50	100	90	26	20	20	18
Turkmeniya	3,770	3,909	4,323	5,534	6,038	6,943	7,023
Ukraine	3,946	3,949	3,436	4,211	3,868	3,769	3,660
Uzbekistan	5,428	7,669	8,541	7,823	8,018	8,034	7,419
Total FSU	359,380	350,619	348,953	348,739	358,490	362,348	390,361

TABLE 1–6
Horizontal Drilling Activity in Russia (1990–2001)

	Wells Drilled	Total Number of Wells Completed
1990	14	14
1991	39	52
1992	50	108
1993	43	151
1994	68	219
1995	91	310
1996	102	412
1997	114	526
1998	128	654
1999	143	797
2000	198	995
2001 (first 9 months)	135	1127

TABLE 1–7
Drilling Volumes in Russia (1994–2000)

	Footage drilled (ft)			Number of Wells		
Year	Exploratory	Development	Total	Exploratory	Development	Total
1994	3,831,040	39,500,000	43,331,040	no data available	no data available	no data available
1995	3,622,924	34,508,360	38,131,284	no data available	no data available	no data available
1996	3,600,000	23,900,000	27,500,000	no data available	no data available	no data available
1997	no data available	no data available	28,990,000	no data available	no data available	4,460
1998	2,590,000	14,140,000	16,730,000	400	2,175	2,575
1999	4,100,000	16,600,000	20,700,000	no data available	no data available	3,185
2000	3,330,000	27,180,000	30,510,000	no data available	no data available	3,405

Value of Scientific Research and Design and the Feasibility of Their Use in Modern Drilling Practices

The successful development of vocational training for industry specialists and the effective work of scientific and research organizations have created a comprehensive engineering knowledge base for developing the oil industry as a whole and drilling technology in particular. For several decades, thousands of engineers and researchers worked on improving drilling technology and equipment. The industry saw the successful development and application of various new technologies, such as the HDHM and EDM, directional and clustered well drilling, etc. Yet, the work on further improvements in the drilling process has slowed.

At the same time, the number of the new technologies developed, tested and prepared for commercial application by the scientific research institutes and design bureaus increased through the years. This led to some paradoxical situations where new technological processes and equipment that were successfully tested by the industry were not used by the companies or promoted by the Ministry officials. Meanwhile, unique developments related to oil and gas drilling accumulated during several decades, and hundreds of inventions represented a significant asset of drilling science. Unfortunately, that knowledge has never found its way into the industry despite its great potential.

A few more examples illustrate these new technologies. Among them were such innovative techniques as dual-bore drilling, i.e., drilling two wells simultaneously using one rig and one drilling crew. This technology was discontinued by order of the Central Committee of the U.S.S.R. Communist Party because of fierce disputes on the work financing issue. Other developments included air, gas, and foam drilling, and casing drilling using retrievable drilling tools.

The Iron Curtain existed for many years around the FSU. Not only did it block the free flow of information from the West, but it also put an even bigger obstacle in the way of informing Western engineers about achievements in the FSU. For example, only in the last few years were Western specialists able to learn about such achievements as the high level of electrodrilling technology development, successful utilization of the ADP, and wide application of the cluster well drilling method. Even more, they were not aware of the developments that proved efficient when used in the pilot-commercial drilling but were not widely used by the industry.

Indicative of this situation is information from the Canadian company Tesco Drilling Technologies [24] about the development of casing drilling technology that the company has been carrying out for several years. The information shows that Tesco repeated the development path that had been completed by Russian engineers long ago. Moreover, the design of the retractable drilling tools used for this method is very similar to the one developed in Russia in the 1950s.

The truth is that scientific, technical, cultural, and art achievements cannot be kept within territorial borders. Sooner or later they become the domain of all humankind. The sooner it happens, the better. While paying tribute to the outstanding progress of drilling technology in the United States and other countries with a well-developed oil and gas industry, we believe that these drilling professionals would gain much from learning about some of the achievements of Russian scientists and engineers working in this field.

Why We Decided to Write This Book

From time to time, visiting U.S. specialists expressed interest in the U.S.S.R. oil and gas industry. This was possible during the "warm periods" in the international political situation. In 1974, John Rowley, a prominent drilling engineer, visited Moscow with a group of U.S. oilfield experts. They also went to the Western Siberia cities of Tumen and Nizhnevartovsk as well as Tatariya and Bashkiriya Republics. The results of this visit were presented at a Society of Petroleum Engineers (SPE) conference in Denver in the fall of 1977. The major topics were oilfield development, drilling technology, and equipment. [25] This was perhaps the first look at Russian technology such as turbodrilling, cluster wells, super-deep drilling with aluminum pipe, multi-lobe PDM, etc.

A delegation from the U.S. Department of Energy (DOE) visited the U.S.S.R. during perestroika. This was actually the first time that the authors of this book could directly discuss different technology features with specialists from the United States. Most of the subjects mentioned in this chapter were presented at a workshop in the VNIIBT followed by informal discussion. At the time, it became clear that the DOE had an interest in turbodrilling and other developments, but it was clear also that there was a great lack of information between the parties.

The idea of writing this book became a relevant question more than a decade ago when Western oil and gas specialists began looking at the FSU as a potential market for exploration and development opportunities. Strong interest led the authors to give a presentation to the Petroleum Industry Forum concerning the "Features of Russian Drilling Technology Development," as arranged by the British Geological Survey in Edinburgh (October 1993). Additionally, in Paris (1995), we held a workshop for the Institute Francaise du Petrole on the same topic, attracting a great deal of interest. This experience strengthened our resolve to write the book.

Since 1995, one of the authors (M. Ya. Gelfgat) has participated in several conferences outside Russia arranged by the SPE, International Association of

Drilling Contractors (IADC), and American Society of Mechanical Engineers (ASME) to introduce the advantages of Russian drilling technologies to Western engineers. [26] [27] From these presentations, it became apparent that the subject of the book would greatly benefit the oil and gas industry, especially since knowledge of the innovative technologies remained closed to the world.

We would like to add that certain Russian technologies have already received varying levels of support for future R&D in the United States, based on cooperative activities involving Maurer Engineering, Inc. (MEI) and the DOE. [28] These joint projects, held from 1996 to 1998, resulted in the DOE targeting several Russian advanced technologies for potential commercial development. Among them were ones we have already mentioned: electrodrilling, ADP, RBs, and reduction gear turbodrills. Industry interest in these subjects is exhibited by the attendance of key operating and managerial personnel at the demonstrations held at MEI in 1999 on RB and ADP technologies. [29] As these and other technologies are used on a commercial basis in the West, this book will serve to answer technical questions that are certain to arise.

During the workshops, seminars, and industrial meetings held during the last few years, the authors received additional confidence that problems in modern drilling practices such as hard- and/or hot-rock drilling, borehole walls instability, super-deep/long drilling, drilling rate improvement, development of complicated fields, and others may be resolved better by understanding the experiences presented in this book.

References

[1] Lisichkin, S. M., *Overview of the Domestic Oil Industry Development* (before the revolution time), Gostoptekhizdat, Moscow-Leningrad, 1954.

[2] Dinkov, V. A., "Petroleum Industry," *Mining Encyclopaedia*, Moscow, Sovetskaya Encyclopaedia, v.3 pp 475–478, 1987.

[3] Lisichkin, S. M., 1954.

[4] Dinkov, V. A., 1987.

[5] Shatsov, N.I., "Introduction," *Oilwell Drilling Handbook*, edited by M.A. Evseenko, Gostoptekhizdat, Moscow-Leningrad, 1947.

[6] Ibid.

[7] Gelfgat, Y.A., "On the history of directional wells drilling and oilfield development with cluster drilling techniques using turbodrills in the U.S.S.R.," *Of the U.S.S.R. Oil and Gas Induty History–Veterans memoirs*, Issue 1, pp. 37–43, Moscow, VNIIOENG, 1991.

[8] Shumilov, P.P., *Theory of Turbodrilling*, Gostoptekhizdat, Moscow-Leningrad, 1943.

[9] Shatsov, N.I., 1947.

[10] Ioanessyan, R.A., *Turbodrilling of Vertical and Directional Wells*, Gostoptekhizdat, Moscow, 1945

[11] Bronzov, A.S., *Cluster Wells Construction at Oil and Gas Fields*, Gostoptekhizdat, Moscow, 1962.

[12] Grigoryan, A.M., *Drilling-in with Multilateral and Horizontal Wells*, Nedra Publishing, Moscow, 1969.

[13] Gelfgat, Y.A., *Drilling-out Cementing Plugs with Turbodrill*, Gostoptekhizdat, Moscow-Leningrad, 1949.

[14] Lisichkin, S.M., *Overview of the Oil Production Industry in the U.S.S.R.*, U.S.S.R. Academy of Sciences Publishing, Moscow, 1958.

[15] Ibid.

[16] Ibid.

[17] Production Catalog of OAO NPO Burovaya Tekhnika, VNIIBT, Moscow, 2001.

[18] "Oil and Gas Industry Status," *World Oil,* v.212, August 1991.

[19] "FSU/Eastern Europe Report," *World Oil,* August, 1996, 1997, 1998, 1999, 2000, 2001.

[20] Gaddy D.E., "Pioneering work, economic factors provide insights into Russian drilling technology," *Oil & Gas Journal,* July 6, 1998.

[21] Gaddy D.E., "Russian test facility allows direct inspection of bottomhole assemblies under in situ conditions," *Oil & Gas Journal,* December 7, 1998.

[22] Gaddy D.E., "Russia shares technical know-how with U.S.," *Oil & Gas Journal,* March 8, 1999.

[23] Bureniye (Drilling) – Special attachment to *Neft I Kapital (Oil & Capital)* #2, November, 2001.

[24] Tessari, Bob, Garret Madell, and Tommy Warren, "Drilling with casing promises major benefits," (Tesco Drilling Technology), *Oil & Gas Journal,* May 17, 1999.

[25] Rowley, John and J. Wade Watkins, "A Tour of Russian Oilfield Technology," *SPE 6718,* 52nd Annual Fall Technical Conference, Denver, Colorado, USA, 1977.

[26] Mnatsakanov, A. V., M. Ya. Gelfgat, and R. S. Alikin, "Technology and technique for scientific drilling in crystalline rocks: experience and perspectives," *IADC/SPE paper #023912,* New Orleans, Louisiana, USA, February 17–21, 1992.

[27] Gelfgat, M.Y., Papers presented at ASME Petroleum Division Drilling Symposiums in Houston: *"Drilling Tools for Continuous Offshore Operations" (1995); "Retractable Bits Development and Application" (1996); "Hydraulic Hammer Drilling Technology: Developments and Capabilities" (1997); "Aluminum Tubular In Deep Water Drilling Application" (1999).*

[28] Maurer W., "Russian Drilling Technologies," Natural Gas Conference (U.S. DOE), Houston, USA, 1997.

[29] Gaddy D.E., "Russia shares technical know-how with U.S.," *Oil & Gas Journal,* March 8, 1999.

2

DOWNHOLE MOTOR DRILLING TECHNOLOGY AND APPLICATIONS

Development of Turbodrills—Characteristics and Fields of Application

From a gear-reduction turbodrill with a single-stage turbine to a multistage hydro-turbine motor

Initial tests. The first turbodrill capable of operation was developed and built by the engineers M. A. Kapelyushnikov, S. M. Volok, and M. A. Kornev in 1922 and 1923.[1] The first prototype was tested by drilling various rock blocks at the surface. The prototype was essentially a single-stage 3 to 4 horsepower turbine with a rotational speed of 1600–1800 rpm. A special hydroturbine reduction gear was used to transfer the rotational effect to a drillbit.

The first tests showed positive results, which enabled manufacturing a small series of turbodrills (8 in. and 11 in. diameter) at the metallurgical plant in Leningrad. In 1925 and 1926, drilling departments of the Azneft production company in Baku began using these turbodrills for drilling wells in the oilfields of the Surakhany region. From 1925 to 1934, the Kapelyushnikov-designed turbodrills that were used to drill nearly 100,000 m in wells with depths of 670–1300 m.

During this period, significant improvements were introduced to the turbodrill design, such as replacement of the reaction turbine with the impulse turbine and replacement of a reducing device with a multistage planetary gear, which was developed by engineer B. G. Lyubimov. The reduction gear and bearings lubricating system design were also improved. Using specially shaped blades for the guide and turbine wheels and improved quality metal and heat treatment of the turbodrill components increased the turbine efficiency factor. Figure 2–1 shows the turbodrill design with the multistage reduction gear. [2]

Tables 2–1 and 2–2 present the results of the turbodrills' annual performance and their comparison with rotary drilling results. These data clearly indicate that in the second half of the 1920s, the overall drilling rate increased satisfactorily using turbodrills. By the early 1930s, it doubled in comparison to the initial drilling rate and then became more or less stable. Nevertheless, the results of turbodrill performance were so much lower, compared to the rotary technique, that further use of turbodrills was considered uneconomical. However, the very fact that turbodrills were used in commercial drilling, even in wells that were considered deep at that time, was of great significance.

TABLE 2–1
Performance of M. A. Kapelyushnikov Turbodrill (TD)

Years	Number of Wells	Overall Drilling Footage, m	Drilling rate, m/rig-month
1925–1926	10	1,324	39.5
1926–1927	16	4,073	45.5
1927–1928	26	6,579	57.6
1928–1929	32	9,714	62.2
1929–1930	49	12,372	65.0
1931	34	12,250	57.0
1932	39	18,837	81.0
1933	24	8,731	78.0
1934 (first half)	16	3,415	72.2

TABLE 2–2
First Turbodrilling Comparison with the Rotary Drilling Result

	1932		1933		1934 (first half)	
	TD	Rotor	TD	Rotor	TD	Rotor
Drilling rate, m/rig-month	82.4	239.8	67.9	134.8	68.1	169.5
ROP, m/hr	0.68	2.2	0.64	1.29	0.67	0.94
Drilling time, hr	4.3	5.9	4.6	9.9	4.0	7.0
Footage per bit, m	2.92	12.98	2.94	12.78	2.68	7.58
Average well depth, m	763	863	1,018	1,000	1,300	1,081
Cost per meter, rubles	187	56	271	106	404	101

In the second half of the 1920s, when the application of turbodrilling technology had made good progress, some American companies showed an interest in this drilling method. M. A. Kapelyushnikov was invited to the United States to perform demonstration test drilling of wells using a turbodrill. Kapelyushnikov came with a drilling crew that had experience in turbodrill application. However, the drilling was discontinued after the inventor of the turbodrill was injured in a car accident. After a long medical treatment in the United States, he returned to the FSU.

The main reason for unsatisfactory performance results from the turbodrill designed by Kapelyushnikov was the low durability of a single-stage turbine. Adequate power of 10–12 hp could be achieved by increasing the drilling mud circulation rate, which provided the turbine wheel rotational speed of 3000 rpm at a drilling mud flow velocity of 100 m/sec. At such velocities, turbine blades were affected by intense erosion, and the maximum durability was as low as 3–5 hr. After that short operating period, the turbodrill had to be replaced. In addition, the wear of the turbine fluid during drilling significantly reduced the power and slowed down the penetration rate during a single bit run.

Another weak point of the turbodrill design was the multistage reduction gear system with a gear ratio from 7 to 150. Rapid wear on the system's components, especially when they were affected by drilling mud, also resulted in the early pullout of a drillstring (DS) for turbodrill replacement. These fundamental drawbacks of the turbodrill design prompted suspension of further development of a single-stage turbodrill.

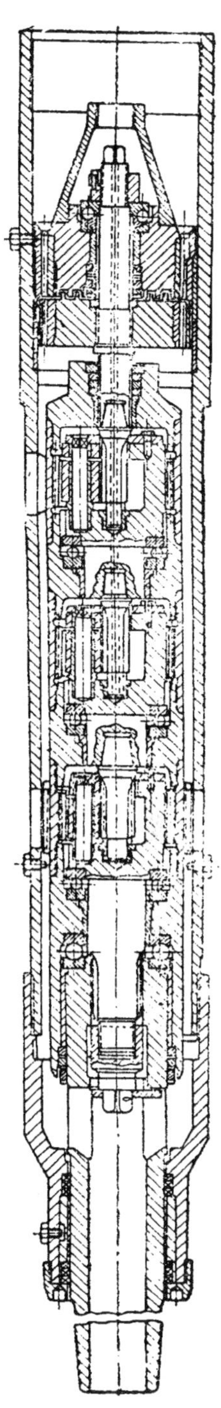

Fig. 2–1 Turbodrill of Kapelyushnikov with three-stage reduction gear (1925–34)

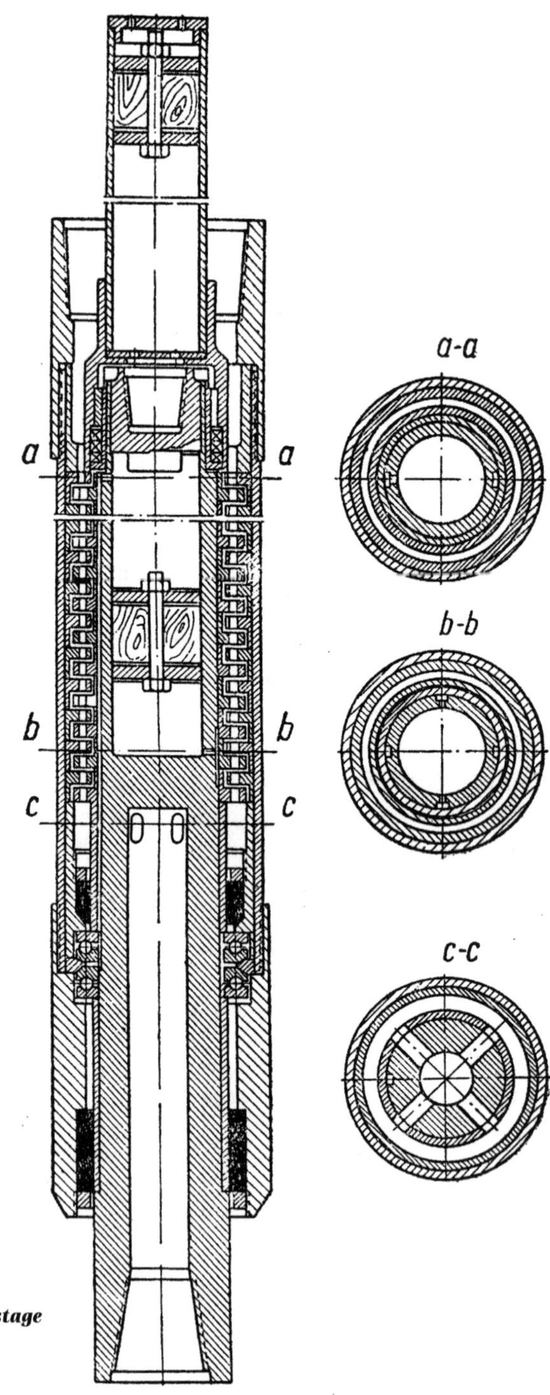

Fig. 2–2 A Direct-drive 100-stage not geared 12-in. turbodrill (1934–1935)

The new engineering team conducted further studies and development work related to turbodrilling technology. This team included P. P. Shumilov, a young scientist from the State Research Institute of Oil (GINI) in Moscow, and E. I. Tagiyev, R. A. Ioannesyan, and M. T. Gusman, young engineers from the State Research Institute of Oil Engineering (GIINMash), along with Lyubimov, an experienced engineer from the same institute who worked closely with Kapelyushnikov. Their joint efforts initially resulted in the development of a 6- to 8-stage turbodrill with a reduction gear system. The turbine had significantly higher power, compared to a single-stage design, and featured much lower wear intensity by the fluid flows. However, the reduction gear system turned out to be the weak link of this high-powered turbodrill design and did not obtain acceptable results.

New multistage design. Since the geared turbodrill idea failed to be reliable, engineers Shumilov, Tagiyev, Ioannesyan, and Gusman from the turbodrilling bureau in Azneft joined efforts in 1934 and 1935 in developing a 100-stage direct drive 12-in. turbodrill (*see* Fig. 2–2). [3]

A thread connection on the turbodrill nipple jointed to the body and compressed the group of stators; friction held the turbine stator and rotor and prevented them from turning inside the housing and on the shaft. A special nut was used to compress the rotors together on the shaft. In addition, they had special slots that were used to set them on a key that was fixed in a slot along the entire length of the turbodrill shaft. The turbodrill design featured an oil-lubricated axial ball bearing located in the lower shaft section under the turbine, as well as a lubricated radial centering roller bearing on the upper shaft end. The internal shaft space contained lubricant that was squeezed out by a special piston. Special seals prevented mud from getting to the inside space of the lower ball bearing and the upper friction bearing.

Power for this new design turbodrill was about 100 horsepower at a shaft rotational speed of 600 rpm. In 1935 and 1936, a small series of these motors were tested in Baku during the oilfield development of the Kaganovichneft Production Company. The turbodrills proved to be efficient and showed better results, such as higher penetration rates, compared to the turbodrills with gear reducers (with low number of stages).

Table 2–3 illustrates the comparison between the results of the multistage gear reduction turbodrills (8 stages) and the first multistage (100 stages) direct drive turbodrill.

TABLE 2–3
The Comparison of Gear Reduction and Regular Turbodrills Performance at Kaganovichneft Company (1935–1936)

	Multistage Gear Reduction Turbodrill	Multistage Direct-drive Turbodrill
RPM	150	600
Power on the bit (HP)	57	92
Torque (kg m)	275	110
ROP (m/hr)	4.6	7.4
Average performance per run (m)	39	40
Drill time (hr)	8.5	5.4

The test results allowed the following main conclusions:

- erosion wear of the turbine fluid paths was not observed in all the stages
- durability of the turbines was no longer a restricting factor for the drilling progress

The tests pioneered the successful application of RKh drag bits at a rotational speed of 600 rpm. Bit weights were reduced because of the low torque of the turbodrill and high torque power of the RKh bit.

Contraction of the stators and rotors using thread connections proved to be quite effective. The bearings turned out to be a weak link in the turbodrill design. Failure of both the lower axial ball bearing because of the poor lubricating system design or the upper radial friction bearing caused low turbodrill serviceability. This was especially true of the lower bearing, which had to be replaced as frequently as every 10–12 hours.

The new turbodrill designs featured the innovative rubber-metal bearing. The lower turbodrill nipple that served to compress stators of all stages was bored from the inside. The inside surface of the bored-out space was rubber faced by vulcanization. This rubber face was in contact with the special polished surface bushing.

The first time the drilling mud lubricated rubber-metal radial bearing was used, it was a complete success. Its durability improved and the service life increased to 100–150 hours. The design used axial ball bearing in the upper turbodrill section installed inside the oil lubrication chamber, which also allowed the use of standard industrially manufactured rather than the backyard-made ball bearings.

Using the work described previously, the first multistage turbodrill T6-150-9¾" (Fig.2–3) was built. This turbodrill could be manufactured by the industry and used to drill wells. The turbodrill design featured 150 turbine stages. The turbine itself is built in the form of a circulation vane cascade with straight blades. The rotary disc body was milled to make the straight turbine blades. Next, steel rims were fit on the heated blades. This was a forced choice because the level of casting technology in Baku at that time did not allow steel casting of rotors with formed blades. In the case of a straight vane cascade design, the stage length was as low as 25–36 mm, which enabled installation of 150 stages inside the turbodrill casing. The T6-150-9¾" turbodrill was used by the drilling industry in 1938 and 1939.

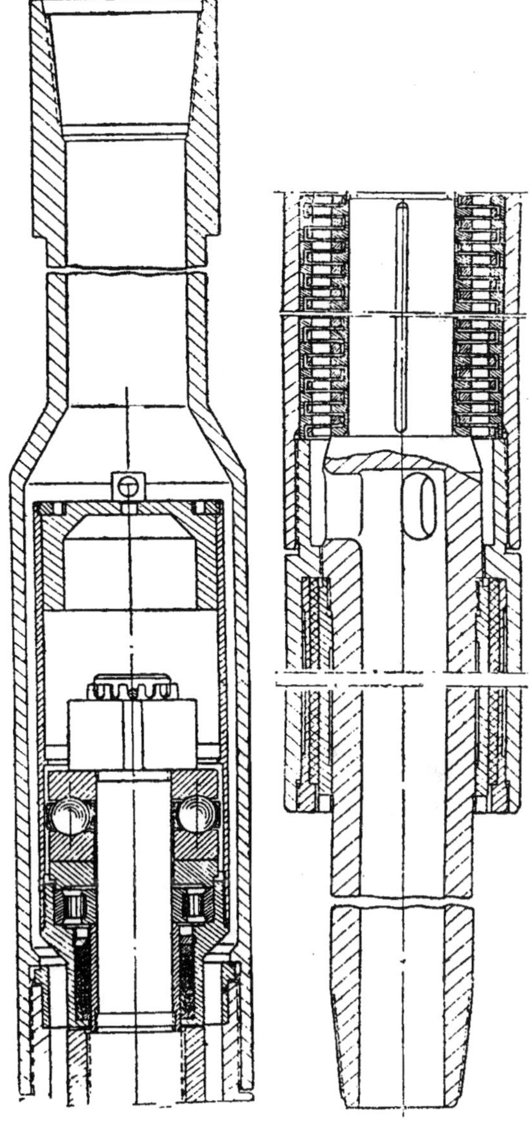

Fig. 2–3 First commercial multistage turbodrill T6-150-9¾" (1935–36)

Drillbits for turbodrilling—first approach. By this time, the rock cone bits found their application along with the widely used cutting-shearing type bits. These new bits proved to be especially efficient in turbodrilling, considering the fact that, for example, the combination of the turbodrill and RKh-type bit was not good in rocks with medium hardness. At high rotational speeds, these bits tended to wear under gage quickly, which led to a much slower ROP and made turbodrilling uneconomical.

Neither satisfied the turbodrilling requirements met by some other bit types, such as the disc bit, the Zublin-design bit, and the FD-type four-point roller cone bit. That was due to the fact that the kinematics of these bits featured intensive cone sliding on the bottomhole, which required high torque and failed to comply with the high rotational speed of a turbodrill. Therefore, engineers from the Experimental Turbodrilling Bureau (EKTB), which was formed in 1939 under the special ordinance from the FSU government, pioneered development of tri-cone bits with conical-shaped cones that provided slide-free or almost slide-free bottomhole cone rolling.

Unlike the normal drilling bureaus, this bureau included an efficient design department, an experimental production facility for manufacturing turbodrills, and other related equipment, including a special drilling test bench for testing new turbodrill designs. The bureau was given the right to lease drilling rigs with crews anywhere in Azerbaijan in order to perform field tests of turbodrills. The design group of the bureau developed this type of bit, fabricated it, and tested it. Later, two specialized plants—the Dzerzhinsky plant in Baku and the Verkhne-Sergiyevsky plant in the Ural region—began manufacturing this type of bit.

Thanks to the low torque power of tri-cone bits with conical-shaped cones, their application allowed the introduction of significant changes to turbodrilling practices by increasing bit weight, which contributed much to the improvement of the penetration rate. In the following years, the tri-cone drillbit became the main tool for destroying rock used by the drilling industry all around the world, including the United States and other countries. At the same time, this innovative tool required the introduction of substantial changes to the turbodrill design.

First design improvements. Several things demanded turbodrill design changes, for example, failure of the axial ball bearing to comply with the drilling conditions. The bearing was designed to withstand the downward load, since the increased bit weight often overcame the hydraulic load. This suggested the necessity of creating a double-thrust bearing capable of withstanding upward bottomhole reaction loads. The experience also revealed that the bit weight increase resulted in buckling of the shaft, which was indicated by the intensive wear of the rotor disc rims as well as by the turbine blade failures. This prompted the use of the intermediate radial rubber-metal bearing that had a design similar to the lower radial bearing. These changes led to the development of a new T10-100-9¾" turbodrill design, shown in Figure 2–4. [4]

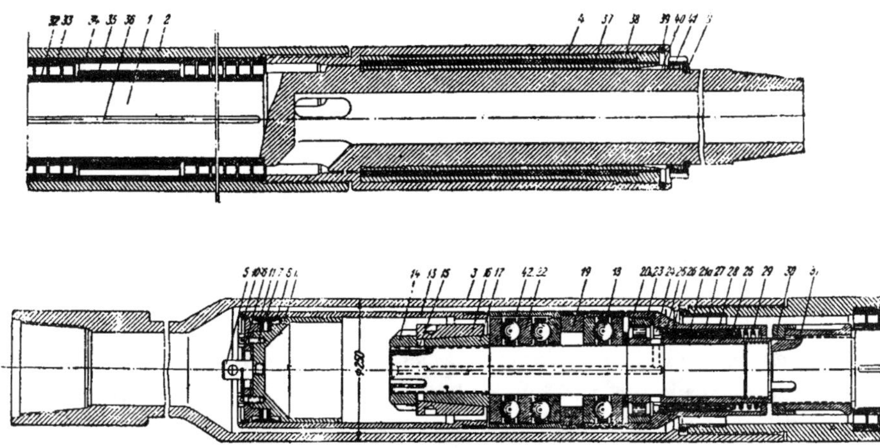

Fig. 2–4 The T10-100-9¾" turbodrill was in serial production in 1940–48

The industry manufactured this turbodrill design in 1940, and it was used by such companies as Azneft in Azerbaijan, Ishimbaineft in Bashkiriya, Krasnokamskneft in the Perm region, and Grozneft in Chechnya. Despite design improvements, the axial ball bearing remained the weak link of the turbodrill design. In this regard, the creators of the turbodrill came up with an idea to use a rubber-metal axial bearing that was quite unique for the world turbodrill manufacturing industry considering its very small diameter.

In 1941, the engineers built the rubber coated turbodrill T12- 9¾". It featured an original design, shown in Figure 2–5. The turbine rotor, consisting of 100 discs, was essentially a collar thrust bearing. The stator was made in the form of a rubber-coated collar thrust bearing, i.e., an axial and simultaneously radial journal bearing because both the end surface and the internal radial surface of the stators were rubber coated. The simple and elegant design solution was quite appealing, especially considering the significantly longer mean time between failures (MTBF) of the rubber-metal axial journal bearing compared to the ball bearing, which was shown by the very first tests in Baku.

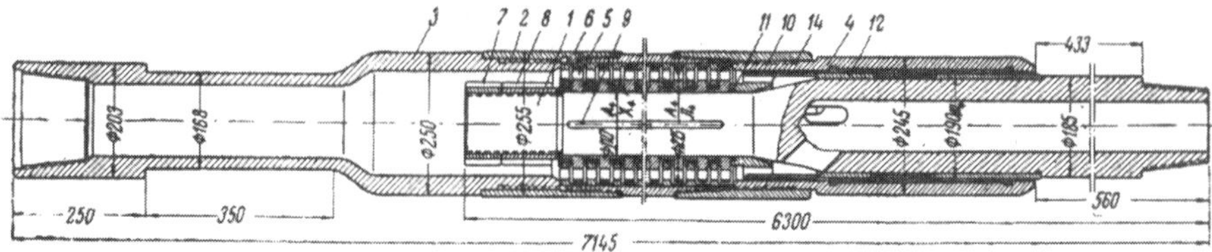

Fig. 2–5 Turbodrill T12-9¾" with rubber-coated stator discs (1941)

At the same time, the tests revealed a number of serious disadvantages related to the combined functions of the motor and the support elements. The designers failed to achieve an even distribution of axial load on the stator along the entire turbodrill length. The support elements of the upper turbine end were the first to wear, which did not allow using the full potential of the multistage support. When replacement of the worn-out elements was required, the entire turbine had to be removed from inside the housing, which complicated execution of the routine repair at the rig. Worse still, the material that was used for manufacturing the rotor disc had low durability because of insufficient abrasive strength and could not be used for this purpose. Exposed to significant vibration loads while drilling through hard rock, the rotor discs sheared the key and began turning on the shaft.

It was then that the engineers came up with the design of the collar step rubber-metal bearing separate from the turbine and located at the upper end of the turbodrill, similar to the T-10 turbodrill design but without the lubricating system. In this design, a drilling mud flow served to lubricate the bearing. The design featured step bearings with special mud courses for lubricating the surface and the heat-treated steel discs with smooth surfaces located between the step bearings. The latter was in the form of 8 to 10 rings abutted against the turbodrill housing. It was gripped by its upper sub and attached to the housing using a straight thread connection. The discs, together with the spacer rings that were used as the upper radial bearing, were fit on the upper end of the turbodrill shaft and held in position by a nut and a lock nut.

Figure 2–6 shows the T14-9¾" turbodrill design with the upper rubber-metal bearing. Figure 2–7 shows the rubber-coated step bearing design with 12 fluid courses. The T14 turbodrill was the last model of a HDHM, which the industry started manufacturing before 1941. During World War II and in the postwar period through 1948, T10 and T14 turbodrills remained the most used models and were manufactured by the industry to satisfy the needs of turbodrilling carried out in some regions of the FSU.

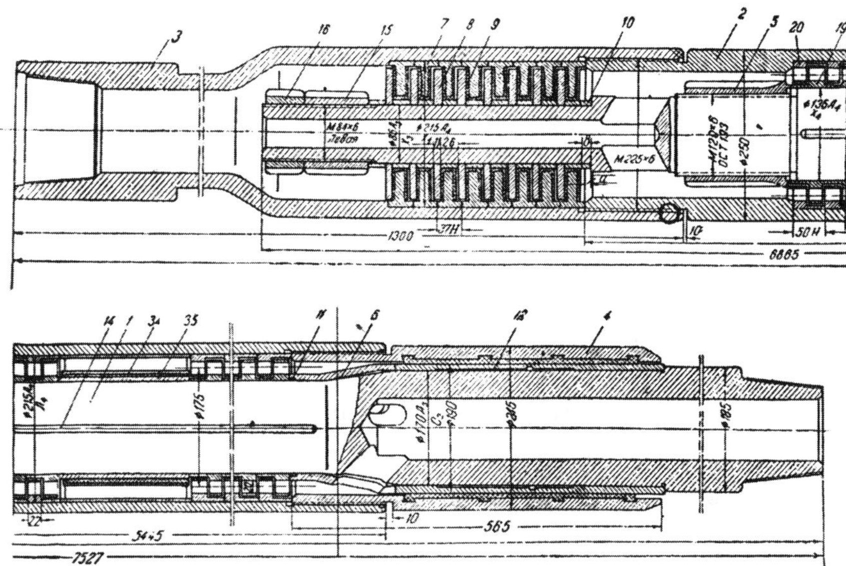

Fig. 2–6 The T14-$9\frac{3}{4}$" turbodrill with the upper rubber-metal bearing was in serial production in 1941–1948

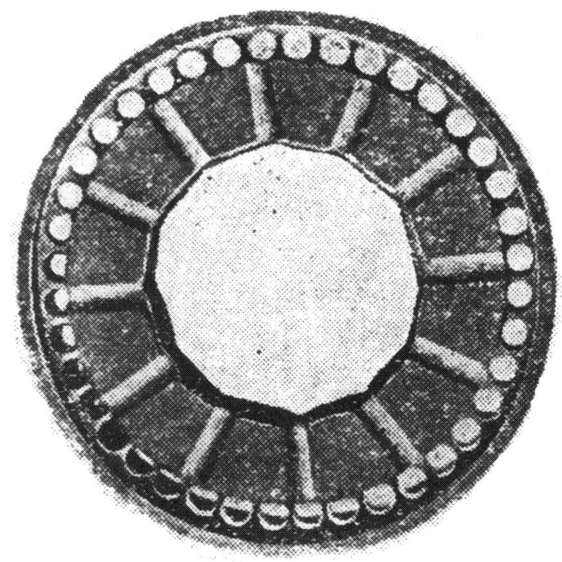

Fig. 2–7 Rubber-coated bearing disc with 12 fluid courses part of axial rubber-metal bearing in T14 turbodrill

Development of the multistage turbine theory

Background. During the period previously described, active work to improve the turbodrill design went parallel with the research work of the group led by Shumilov to develop the multistage turbodrill theory and design components. Shumilov set forth the main provisions of this theory in his works *Oil Well Turbodrilling,* Volume I and II, published in 1936 [5] and *Fundamentals of Turbodrilling Theory,* published in 1943. [6] Tragically, the latter was the last book by Shumilov, who died in 1942. In 1968 his sons, V. P. Shumilov and L. P. Shumilov, who were oilfield industry engineers also, published the book *Oil Well Turbodrilling; Selected Works,* [7] which was essentially a collection of the elder Shumilov's works. The book aimed at making primary sources available to a new generation of engineers since by that time these sources had become rare books.

A brief description of the most important postulates of the multistage turbodrill turbine theory, based on Shumilov's works, is given as follows. Nearly all courses related to turbine theory and designs include brief information on the axial flow (cylindrical) turbines. However, this information is not sufficient for development of a multistage turbine design useful for drilling wells. These turbines must possess such characteristics, as high power, low speed, small flow rates, high pressure, and the power to operate in contaminated flow conditions. Also, they must have a small diameter. These conditions of oil well drilling required development of the special theory and the method of designing axial flow multistage turbine. This theory is based on the following fundamental postulates.

Postulate of identity of individual stage action. A pair of neighboring discs, a guide disc and a working disc, forms a turbine stage. The working discs are fit on the turbine shaft and make the same number of turns. The guide discs are located with certain required clearances between the working discs and are fixed in the stationary turbodrill housing. While designing the hydraulic multistage turbine, the engineers assumed full identity of the action of all sequentially located stages.

The condition of the action and design identity of all stages allowed the basic calculation of a multistage turbine based on the hydraulic and mechanical analysis of one stage.

Postulate of Symmetry of the guide and working vanes. The principle of symmetry in the turbine stator and rotor interaction pattern postulates the following. The field of velocities of the relative flow pattern in the rotor discs fluid course must be a mirror reflection of the field of velocities of the absolute flow pattern in the stator discs courses. If this condition is met, the turbine length can be optimized through installation of the maximum number of possible stages. Using vanes with a mirror reflection pattern fulfills this condition. The rotor disc vane profiles (left-handed vanes) are essentially mirror reflections of the stator guide disc profiles (right-handed vanes). Figure 2–8 shows an example of such a profile. [8]

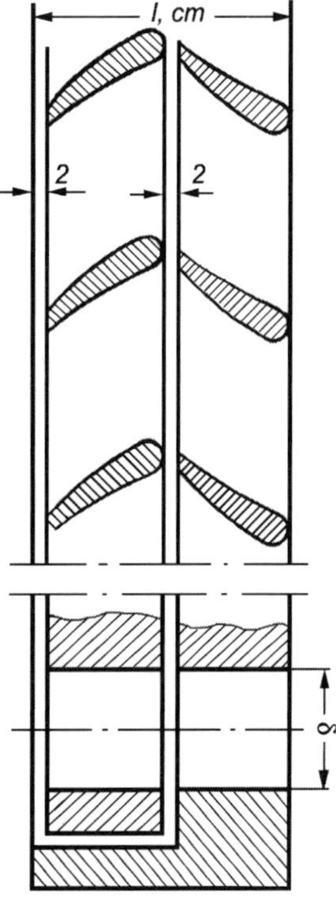

Fig. 2–8 The plain blade rotor/stator discs vane profile

The uniform size theory of axial flow turbine application. The fluid flow through the turbine courses is a complex phenomenon with its parameters varying in time and space. The flow inside the turbine with rotating rotor and regularly blocking stator courses pulsates periodically, which affects the mechanical flow properties. In addition, the plunger mud pump operation pulsates as well. [9] The notion of the constant average value of the flow velocity in the stator and the flow velocity in any other area of the turbine is used here.

This method of the simplification issue regarding both time (disregarding pulsation) and space (the notion of average stream) is the essence of the uniform size theory of turbines.

The postulates specified previously were used to develop the reduced forms of the main Eiler equation of the theory of turbines that is important from a practical standpoint. The analysis of the axial flow turbines used for drilling wells is based on the uniform size theory.

The theory of turbine equations. The fundamental equation of the Eiler theory of turbines can be written in two forms

$$\frac{C_{1u}U_1 - C_{2u}U_2}{g} = H_1 \qquad 2.1$$

$$\frac{C_1^2 - C_2^2}{2g} + \frac{U_1^2 - U_2^2}{2g} + \frac{w_2^2 - w_1^2}{2g} = H_1 \qquad 2.2$$

where

C_1 is the absolute velocity of the stream at the stator outlet and the rotor inlet

C_2 is the absolute velocity of the stream at the rotor outlet and the stator inlet

$w_1 w_2$ are the relative velocities

U_1, U_2 are the peripheral velocities at the turbine rotor inlet and outlet

u is the mean projection onto the peripheral velocity direction

H_1 is effectively used flow pressure in one stage of the turbine

g is the acceleration of gravity

For the axial flow turbines with blades of constant radial length, the peripheral velocities U1 and U2 are equal; therefore, the formula takes the following form:

$$\frac{(C_{1u} - C_{2u})U}{g} = H_1 \qquad 2.3$$

$$\frac{C_1^2 - C_2^2}{2g} + \frac{w_2^2 - w_1^2}{2g} = H_1 \qquad 2.4$$

The axial flow turbine analysis is based on these forms of the Eiler equation. The result H_1 is an effective power transmitted by every kilogram of the workflow to the turbine blades, regardless of the turbine efficiency factor. The effective power, W_1, transmitted to the axial flow blades can be described by the following formula:

$$W_1 = \frac{(C_{1u} - C_{2u})U}{g\gamma} Q \qquad 2.5$$

where

Q is the volumetric flow rate

γ is the fluid specific gravity

Power divided by the angular velocity of the turbine rotation equals effective torque, M_1, applied to the turbine blades.

$$M_1 = \frac{(C_{1u} - C_{2u})r}{g} Q\gamma \qquad 2.6$$

where

r is the radius of the cylindrical surface of the workflow average stream movement

Analyzing the Eiler theory of turbines with regard to the multistage turbines, Shumilov studied the relevance of these equations to the adverse losses of the workflow power. In this connection, he made some important conclusions related to the issue of the effect of the turbine efficiency factor on its effective power. These conclusions stated that additional (hydraulic) losses in actual conditions reduced the effective power and torque because of the lower flow rate in the case of spontaneous workflow through the turbine, which varied depending on the level of resistance forces to be overcome.

In the case of forced flow (the work fluid flow rate that is not dependent on the additional resistance force level), the effective power and torque of the turbine blades do not depend on the losses in the turbine.

In conditions of forced flow, the type of fluid (viscosity level) and the manufacturing imperfections of the blades' work surface affect only the additional pressure loss level and not the realized effective power (and torque).

These comments are of value for turbodrilling that involves operation in forced flow conditions with a fixed rate in the turbine due to its generation by the piston pumps.

The hydraulic and impact pressure losses in the turbine increase its pressure consumption (thus reducing the efficiency factor). However, these losses do not affect the pressure effectively realized at the blades represented by Equations 2.1 and 2.2. The Eiler equations allow singling out the turbine flow mechanics analysis (effective power and torque) from the turbine hydraulics analysis, which involves complex efficiency factor analysis problems.

Vortex type flow in an operating turbine—"degree of circulation." The workflow enters the stator fluid courses carrying certain energy density content (energy per kilogram of fluid). In the beginning of the stator fluid courses, this energy density content mainly takes the form of the hydraulic head. Along the length of the stator course, the hydraulic head gradually transforms into a velocity head because of the reducing cross-section area. At the end of the stator courses, the entire flow is involved in rotation around the turbine axis as a result of the blades turning. At the outflow point, the stator generates a high intensity vortex that has an axis that coincides with the turbine axis. The spinning rotor, taking in the ordered vortex, reduces its intensity level or even generates a counter-vortex.

The flow mechanics theory indicates the possibility of generating a vortex using a certain torque that affects the flow. Conversely, reduction of the vortex intensity or its sign reversal is also related to generations of certain torque. Moving the flow round the turbine axis, the stator blades are affected by the counter-direction reactive torque, which must be dampened by fixing the stator system. Reducing the intensity of the vortex flow around the turbine axis (or even reversing the vortex sign), the rotor blades are affected by a reactive torque reverse to the direction of the vortex flow and dampened in the stator.

Because the rotor is fixed on the turbine shaft and can rotate around the turbine axis, the reactive torque applied to the rotor blades, generated by the decrease of the vortex flow intensity, is capable of performing effective work. In the multistage axial flow turbines, the vortex flow intensity at the rotor outlet equals the vortex flow intensity at the next stage stator inlet. Thus, in this case, the vortex intensity increase in the stator equals the intensity decrease in the rotor. Obviously, the turbine effective power equals the torque generated in the rotor by the decrease of the vortex intensity multiplied by the rotor angular velocity. The higher the amplitude of the vortex intensity fluctuation in the stator and rotor, the higher the reactive (working) torque generated in the rotor, and, consequently, the lower the angular velocity of the turbine rotation can be taken for the same turbine power.

The problem of construction of the slow speed axial flow turbines with high torque can be resolved by generating substantial amplitude of the vortex flow intensity fluctuation in the stator and rotor.

The theory developed by Shumilov allowed the introduction of the term *degree of circulation,* which represents a ratio between dynamic and kinematic factors of power and is described by the equation:

$$\sigma = \frac{(C_{1u} - C_{2u})r}{U} \qquad 2.7$$

The low-speed turbine with a high dynamic factor of power has a high degree of circulation when operating on an optimal (non-impact) regime. In contrast, the high-speed turbine with a low dynamic factor features a low degree of circulation but at an optimal operational regime.

These regularities are very important in selecting the type of turbine that meets turbodrilling requirements.

Selection of the degree of circulation hinges upon the turbine operating conditions and the requirements it must meet. When the workflow pressure is low with the simultaneously high rate and the turbine speed must be high, so the turbine must have a low degree of circulation. Conversely, in low rate and high pressure flow conditions, a high degree of circulation is required to build a slow speed turbine. These are typical turbodrilling conditions.

Figure 2–9 shows the blade degradation diagram, determining the range of σ variation from 0 to ∞, which ensures a variety of turbine parameters. Figure 2–10 presents schematic blade types symmetrical with the five types of high circulation multistage turbines (shown in Figure 2–9) that begin with the ultimate $\sigma = \infty$ and end with straight vanes. Since the angles α_0 and β_2 (the diameter and the radial height of the blade) are constant, the torque M_{st} and the number of revolutions $n_{freerun}$ remain the same for all types.

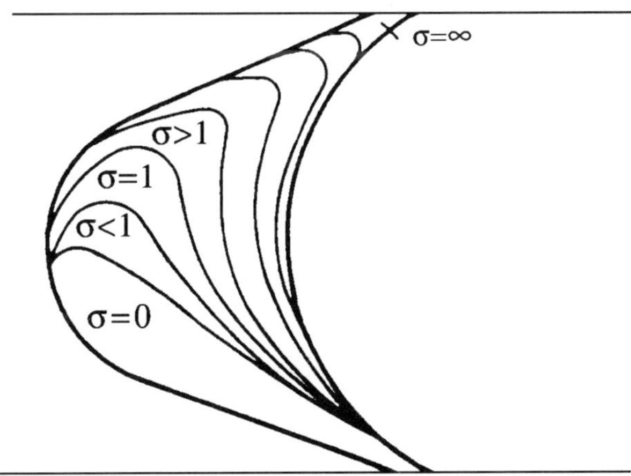

Fig. 2–9 The blade degradation diagram

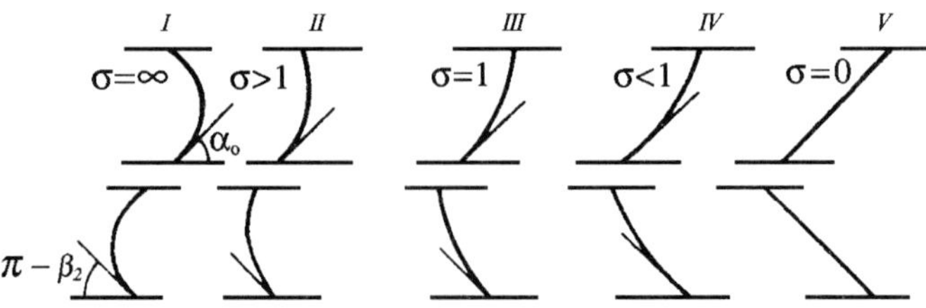

Fig. 2–10 Schematic of five blade types (I, II, III, IV, V) with different degree of circulation applicable for the multistage turbines design

Figure 2–11 presents the diagram pattern, number of revolutions, and efficiency factor (n,η) for the five types of turbines. Figure 2–12 shows the operational pressure vs. rotational speed curve.

Profile No. 2 features slight pressure increases when the operating mode of the turbine is changed from stall to free run point.

Profile No. 3 features an almost constant pressure level.

Profile No. 4 features pressure decrease during the turbine mode changing from the stall to the free run point.

Profile No. 5 generates the highest pressure at the turbine inlet, at constant rate, and at the stall mode, and the lowest pressure at no-load operation.

The extreme types of profiles that featured essential pressure variation when the turbodrill rotational speed increased were attractive because, at a fixed flow rate generated by the piston pumps, these pressure variations indicated the drilling mode of the turbodrill. However, this was the only attractive feature of these profiles since, as mentioned previously, they featured the lowest turbine efficiency factor. True, the low efficiency factor at the fixed flow rate did not affect the turbine power but only increased the hydraulic head (pump pressure). However, the pump power also had its limits. The maximum surge pressure while using No. 1 and No. 5 profiles was a negative factor and resulted in activation of the pump relief valves and interruption of the drilling process.

During the practical implementation of these conclusions, all the profile types were used. The first type, produced by the industry and used in commercial

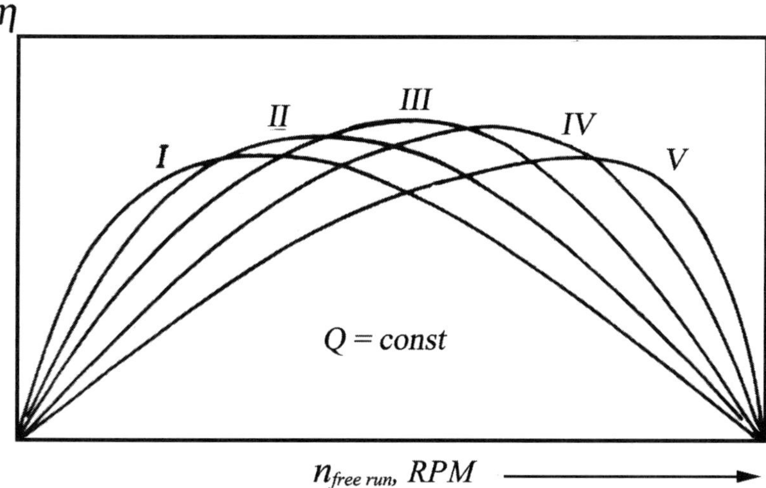

Fig. 2–11 The diagram of efficiency factor (η, %) vs. free run revolutions (n, rpm) for the five different types of turbines with identical flow rate (Q)

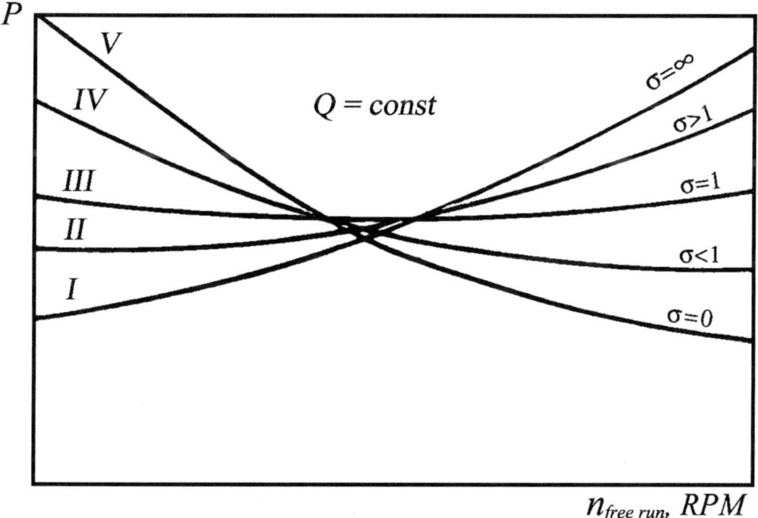

Fig. 2–12 Differential pressure at the turbine stage (p) vs. rotational speed (n) when flow rate (Q) is not changed

drilling, was essentially a vane cascade turbine with straight blades. Shumilov pioneered application of this type of turbine and raised a lot of negative comments from many experts in flow mechanics (regarding the efficiency factor level).

One of them was Academician L. S. Leibenzon, a prominent scholar in the fields of mechanics and hydrodynamics who was a scientific mentor of Shumilov when the latter was a student at the Moscow State University. In 1940, Leibenzon headed a special committee in Narkomneft (the former name of the Russian Oil Ministry) to oversee bench tests of multistage turbines developed by the engineers of the EKTB. The tests aimed at determining the power, efficiency factor, and torque of the turbines. Before the tests, the academician joked that if the efficiency factor of the vane cascade with straight blades exceeded 10–15%, he would take a university course again.

Shumilov was convinced that the efficiency factor of the straight vane cascade would be at an acceptable level. The bench tests proved his belief. Shumilov's desire to use straight vanes could be explained partly by the drawbacks of the existing casting technology, which did not allow production of a high-quality cast blade profile. Still, the main reason for his wish to use these type blades was that due to the shorter stage for the same length of the turbine shaft, they allowed an increase in the number of stages to 150 and, correspondingly, increased the dynamic component of power.

Bench tests of turbodrills and mathematical derivations of power characteristics. The specialists from the EKTB installed a test bench at their drilling rig for test work. The pump unit included two pumps manufactured at the Krasnyi Molot engineering plant in Grozny with a 180 kW induction motor drive. A hydraulic brake was used to measure the characteristics.

Figure 2–13 presents a schematic drawing of the hydraulic brake. [10] The brake consisted of a housing installed on the rotary table and partially filled with water. The impeller with a shaft rotated in the rubber-metal bearings was installed in the housing and connected to the turbine shaft. The turbodrill was positioned vertically and hung on a rotary hook, the turbodrill was moved up and down along its axis using the drawworks. Slowdown of the turbine rotation was achieved by lowering the impeller into the water inside the housing. The housing was filled with water up to a certain level so that it allowed idle rotation of the turbine shaft in the uppermost position.

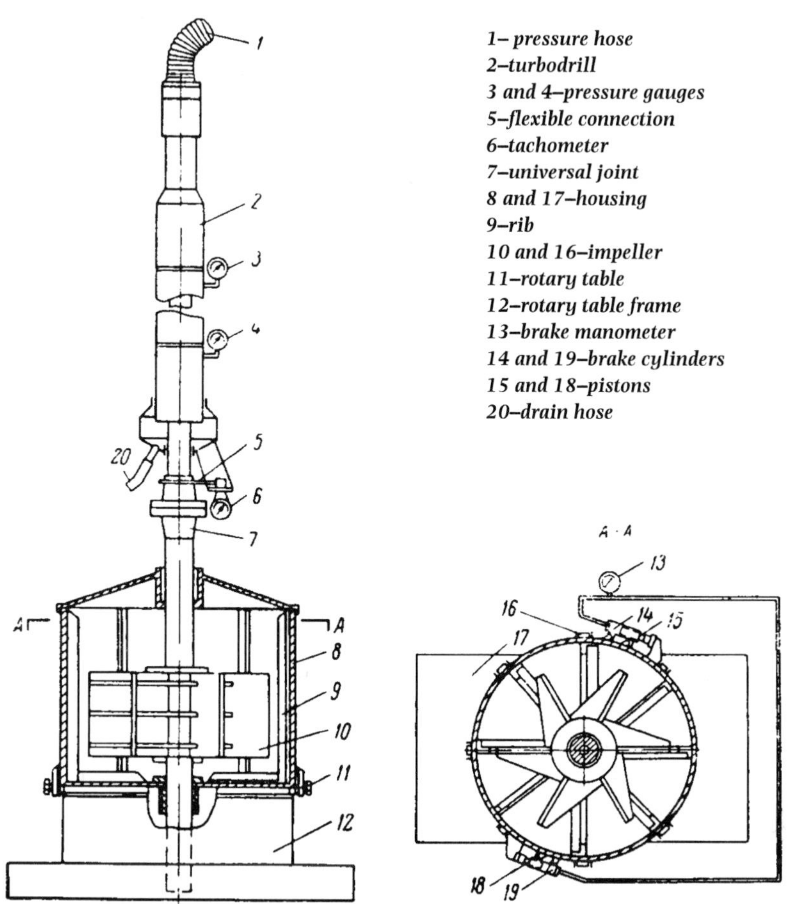

Fig. 2–13 Schematic drawing of the hydraulic brake used at the test facility in Baku for turbines characteristics study (end of 1930s)

The steel guide ropes, attached to the corners of the derrick, dampened the reactive torque of the turbodrill housing. The flaw rate was measured by the volumetric method. It was possible to determine the differential pressure level using the difference in readings of the two pressure gauges installed at the inlet and outlet of the turbine. The torque, applied to the rotary table, was transmitted to the rotary bed using two brake cylinders. The pressure registered in the cylinders has been used to calculate the torque generated by the turbine.

The number of RPM was determined using a tachometer, attached to a sheave wheel that was connected through a flexible joint to the turbine shaft. Test operators, located at each of the instruments previously mentioned, continuously registered the instrument readings. The readings and calculation results were plotted on the board with the following coordinate system: *n* (number of revolutions) plotted on the abscissa, *W* (power), *M* (torque), η (efficiency factor), and *P* (differential pressure) plotted on the ordinate. Figures 2–14, 2–15 [11], 2–16, and 2–17 [12] present some of the resulting curves.

The T6-type turbodrills with straight blades used by the drilling industry were tested first. Next, the same type turbodrill with special profile blades were tested (Fig. 2–15). The T10 type turbodrills with special profile blades only were tested as well (Fig. 2–16 and 2–17).

The tests indicated a sufficiently high efficiency factor of about 60% for the turbines with straight blades. Still, the efficiency factor of the turbines with profiled blades was higher (65%), even considering the low quality of the cast. The blades had a rough and non-streamlined surface. Shumilov presumed that given better quality of the cast turbine discs, the efficiency factor might be higher than 75%. Construction and testing of precision cast and plastic turbines proved this later. Besides determining the efficiency factor, the tests also indicated the possibility of high power generation at lower fluid rates by using a higher differential pressure level in turbines with profiled blades. On the whole, considering the hydraulic losses inside the DS and the borehole, this reached a higher general efficiency factor in the drilling process.

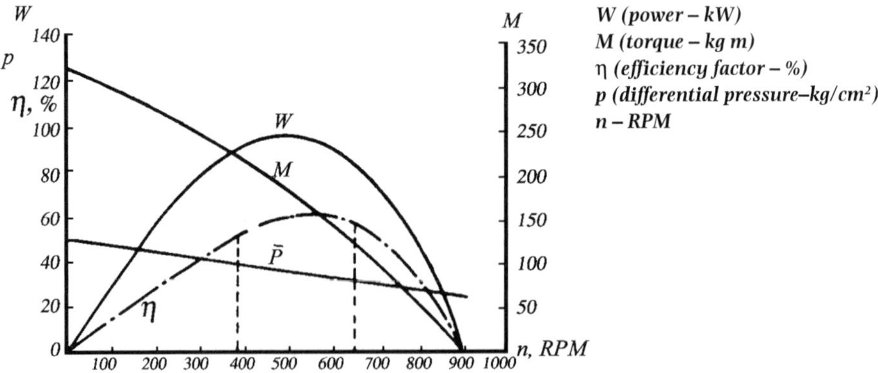

Fig. 2–14 Experimental operational characteristics of commercial turbine T6-150 - 9¾″ 150 turbine stages with straight blades at water flow 45.6 l/sec

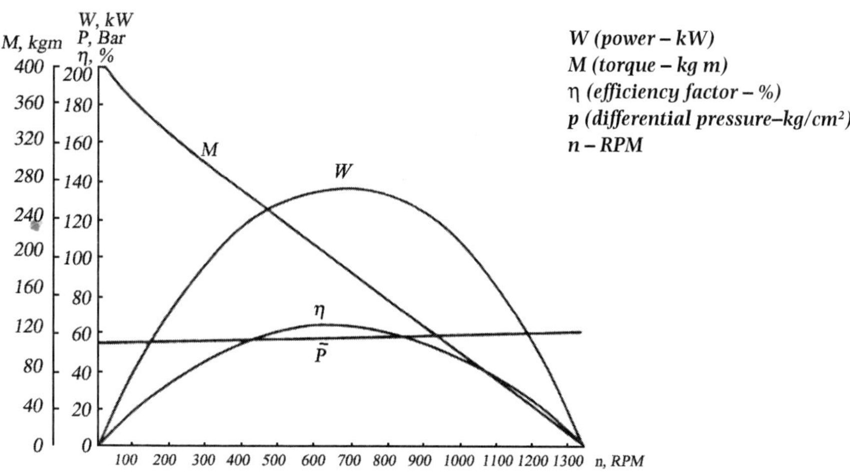

Fig. 2–15 Experimental working characteristics of test turbine T6-100-9¾" 100 turbine stages with profiled blades at water flow 37 l/sec

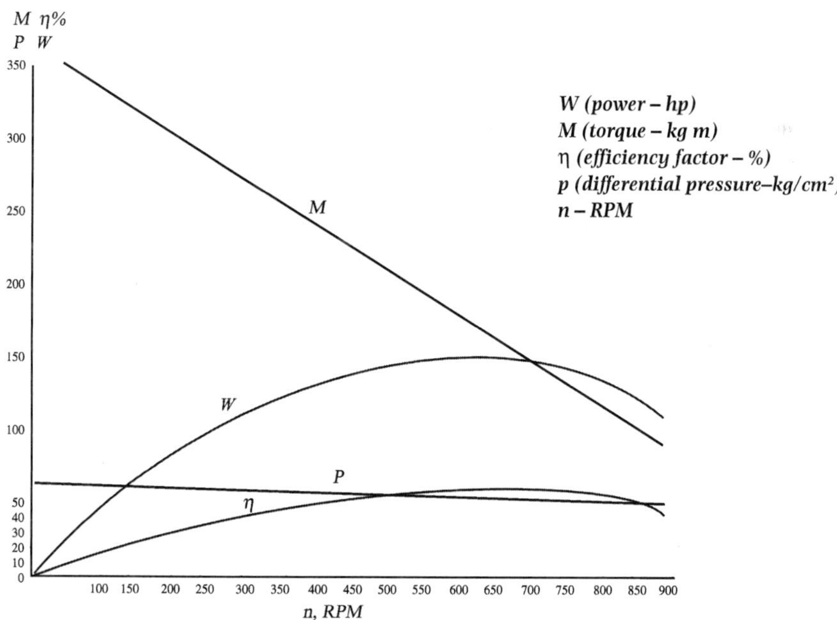

Fig. 2–16 Experimental working characteristics of turbine T10-9¾" at water flow 36 l/sec

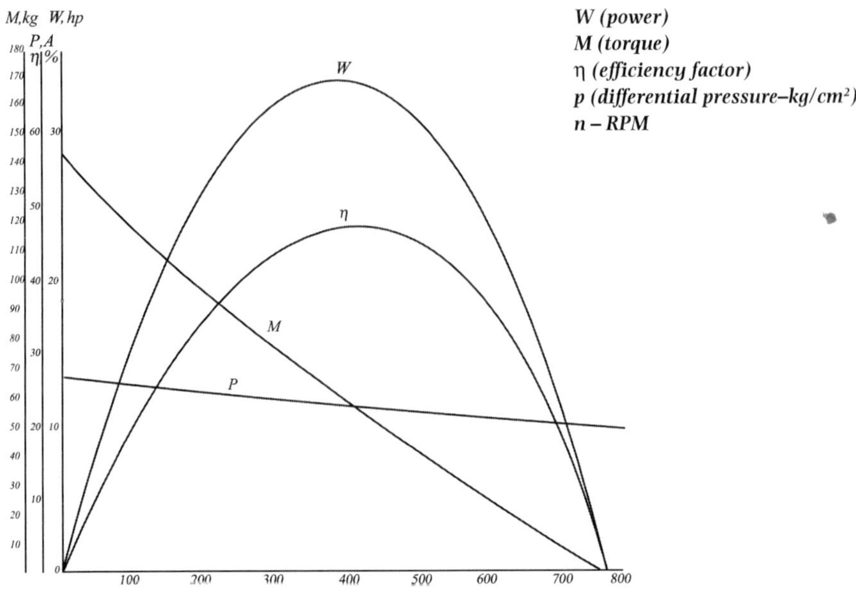

Fig. 2–17 Experimental working characteristics of turbine T10-9¾" at water flow 23,3 l/sec

The conclusions described previously prompted all manufactured turbines to be designed with special profile blades. The majority of blades used were type III and IV (*see* Fig. 2–10). However, attempts to develop a turbodrill with stable characteristics in the braking mode at lower rpm and higher torque resulted in the use of the type "A" turbodrill with type "I" turbines, which are described in detail in the following sections. As Shumilov predicted, this type of turbodrill had a special valve enabling relief of a certain amount of fluid to avoid turbine increased speeds during idle operation.

Regarding the turbodrill bench tests, it should be noted, that in addition to the test bench with the hydraulic brake, a test bench with a steel bottomhole was built to provide the maximum close simulation of the actual turbodrill downhole operating conditions while drilling hard rock. For this purpose, the large size rotary bushing was replaced with a steel borehole simulator (round-shaped steel pig). The weight on bit (WOB) was made by fitting specially molded round pigs, gripped together by long bolts on the turbodrill. The test results built curves showing dependence of the turbine rpm, power, torque, and efficiency factors from the bit weight level.

One of the important features of the test bench with a steel bottomhole and weight applied to the tested turbodrill was the possibility of establishing the dependence between the bit weight level and the torque applied to the turbodrill shaft while using various type drillbits. Using this dependence, the engineers calculated specific torque levels for each bit type, which is important for selection of an optimum bit type.

The theoretical and experimental studies conducted by Shumilov were instrumental in developing the methods for calculating multistage hydraulic turbines and obtaining resultant equations to calculate power, torque, differential pressure, and rotational speed for various design and hydraulic parameters of vane cascades and various fluid rates. Figure 2–18 [13] shows design parameters of the vane cascade with special profile blades.

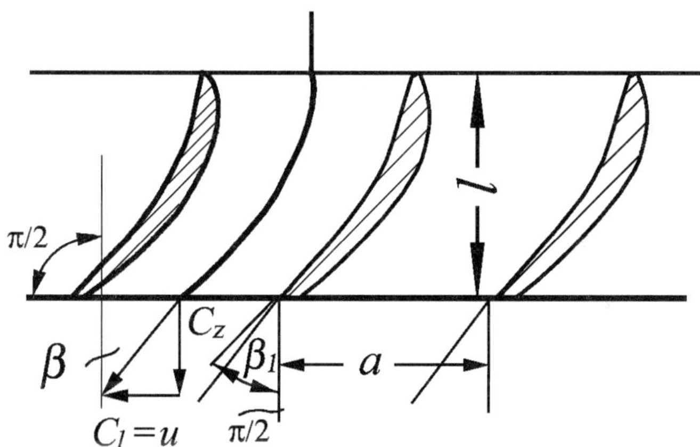

Fig. 2–18 Design parameters of the vane cascade with special profile blades

With the axially directed flow at the stator (absolute motion) and the rotor (relative motion) the level of the velocity work component C_u, included in the Eiler equation, is equal to the blade peripheral velocity u for the average design diameter d of the turbine. Therefore, the Eiler equation for the multistage turbines may take the following form:

$$C_u \bullet u = u^2 = \frac{\eta_m}{\eta_M} \frac{g}{\gamma} \frac{P_m}{K} \qquad 2.8$$

where

P_m is the differential pressure in the turbine, bar

K is the number of stages

η_m and η_M are the full and mechanical efficiency factors of a turbodrill

Following is the equation for the axial component C_z (Fig. 2–18):

$$C_z = \frac{1-\xi}{\chi} \frac{Q}{\pi dh} \qquad 2.9$$

From Figure 2–18, the following equation is derived:

$$u = C_u = C_z \, tg\beta \qquad 2.10$$

Substituting variables u and C_z in Equation 2.8 and resolving it for P_m, we obtain the equation to calculate the differential pressure level in the turbine using the rate and design parameters for the optimum operating mode:

$$P_m = K \frac{\gamma}{g} \frac{\eta_M}{\eta_m} \left(\frac{1-\xi}{\chi} \frac{tg\beta}{\pi dh} Q\right)^2 \qquad 2.11$$

where

ξ is the factor of fluid leak through clearances between the work and the guide discs

χ is the factor of workflow cross-section area constriction in the turbine fluid courses, accounting for blades and zones of turbulent flow around blades

h is the radial width of a fluid course, m

β is the hydro-mechanical angle between the centerline of the flow at the stator outlet in absolute motion (or at the rotor outlet in relative motion) and vertical flow

The formula of the effective power at the bottomhole is derived from the known equation:

$$W_m = \frac{P_m Q}{102}\eta_m \; ; W_m = \frac{k}{102}\frac{\gamma}{g}\eta_{Max}\left(\frac{1-\xi}{\chi}\;\frac{tg\beta}{\pi dh}\right)^2 Q^3 \qquad 2.12$$

The rpm for the turbine is determined from the known equation:

$$n = \frac{60u}{\pi d} \; ; n = 60\frac{1-\xi}{\chi}\;\frac{tg\beta}{(\pi d)^2\eta}\;Q \qquad 2.13$$

Bit torque is calculated using the following equation:

$$M_m = 102\frac{W_m}{\omega} = 102\frac{30}{\pi}\;\frac{W_m}{n} \qquad 2.14a$$

where

ω is angular velocity

Next, variables W_m and n are substituted with the corresponding equations:

$$M_m = K\frac{\gamma}{g}\eta_{Mex}\frac{1-\xi}{\chi}\;\frac{tg\beta}{2\pi d}\;Q^2 \qquad 2.14b$$

The variable β is the angle between the flow stream centerline and the vertical line. To develop a turbine design, the blade design angle β_0 between the vertical line and the tangent to the blade profile at its end point must be known (*see* Fig. 2–18). The blade design angle β_0 is always bigger than the flow angle β because the blade does not fully deviate the stream. Shumilov, using the ideas of Pfleiderer and Spanchake, suggested using the following equations:

$$Sin\beta_0 = \frac{Sin\beta}{1-\frac{\tau}{2}\frac{a}{1}Sin\beta_0} \qquad 2.15$$

$$Sin\beta = \frac{Sin\beta}{1+\frac{\tau}{2}\frac{a}{1}Sin\beta_0} \qquad 2.16$$

where

τ is the coefficient determined from tests and varies between 0.6 and 0.9 mm

l is the blade radial height

$a=\dfrac{\pi d}{z}$ is blade pitch

z is the number of blades in a turbine stage

Equations 2.15 and 2.16 calculate the blade design angle β_0 using the stream deviation angle and vice versa.

After Shumilov had developed these equations for the main power characteristics of turbodrill turbines, he suggested a method to recalculate multistage turbine designs for various operating conditions.

The formulas shown as follows are used to calculate the design parameters for the turbines with special profile blades. These formulas were derived from Equations 2.11–2.14. All primed symbols denote values related to the experimental turbine, whereas the unprimed symbols denote values related to a commercially produced turbine, the characteristics of which are determined using the parameters of the experimental turbine.

$$\frac{W_m}{W'_m}=\frac{K\gamma}{K'\gamma'}=\left(\frac{tg\beta}{tg\beta'}\,\frac{d'h'}{d\bullet h}\right)^2\left(\frac{Q}{Q'}\right)^3 \qquad 2.17$$

$$\frac{M_m}{M'_m}=\frac{K\gamma}{K'\gamma'}=\left(\frac{tg\beta}{tg\beta'}\,\frac{d'h'}{d\bullet h}\right)^2\left(\frac{Q}{Q'}\right)^2 \qquad 2.18$$

$$\frac{P_m}{P'_m}=\frac{K\gamma}{K'\gamma'}=\left(\frac{tg\beta}{tg\beta'}\,\frac{d'h'}{d\bullet h}\right)^2\left(\frac{Q}{Q'}\right)^2 \qquad 2.19$$

$$\frac{n}{n'}=\frac{tg\beta}{tg\beta'}\,\frac{d'^2h'}{d^2\bullet h}\,\frac{Q}{Q'} \qquad 2.20$$

These equations are used to calculate turbine designs for various operating conditions when considering minor changes to the vane design parameters. The equations show the relationship between the main design parameters of the major operating modes. In a particular case of turbine recalculation with changed flow rate, the following equations are used:

$$\frac{W_m}{W'_m} = \left(\frac{Q}{Q'}\right)^3 \qquad 2.21$$

$$\frac{M_m}{M'_m} = \left(\frac{Q}{Q'}\right)^2 \qquad 2.22$$

$$\frac{P_m}{P'_m} = \left(\frac{Q}{Q'}\right)^2 \qquad 2.23$$

$$\frac{n}{n'} = \frac{Q}{Q'} \qquad 2.24$$

The relationship between the values *W*, *M*, and *P* for the optimum slowdown and no-load operating modes for turbines with special blades profile at $\sigma = 1$ are shown as:

$n_{stall} = 0$ $\qquad$ $n_{free} = 2n_{optimal}$

$M_{stall} = 2M_{optimal}$ $\qquad$ $M_{free} = 0$

$W_{stall} = 0$ $\qquad$ $W_{nl} = 0$

$P_{stall} = P_{optimal}$ $\qquad$ $P_{free} = P_{optimal}$

The following shows the relationship between the main power characteristics of various diameter turbines for these types of modes at constant rotational speed. [14]

$$\frac{M_2}{M_1} = \left(\frac{D_2}{D_1}\right)^5 \qquad 2.25$$

$$\frac{P_2}{P_1} = \left(\frac{D_2}{D_1}\right)^2 \qquad 2.26$$

$$\frac{W_2}{W_1} = \left(\frac{D_2}{D_1}\right)^5 \qquad 2.27$$

Earlier in this chapter, the transitional period in the development of turbodrill designs from 1934 to 1941 were described. Table 2–4 presents characteristics of the turbodrills that were commercially produced in 1941. These characteristics were obtained from theoretical studies and bench tests.

TABLE 2–4
Technical Characteristics of the Turbodrills

	Turbodrill Types					
	T10-9¾"	T12-9¾"	T14-9¾"	T14-7¾"	T14-7"	T9-50-5"
Flow rate (l/sec)	40	40	40	30	22	10-12
Effective power (HP)	190	180	189	100	55	10-15
RPM at maximal power	650	600	600	685	665	1,500
Torque (kg.m)	205	205	205	105	60	5-7
Pressure drop at turbine (MPa)	6.83	6.12	6.12	5.61	4.08	3.06
Efficiency (%)	55	50	50	45	40	40
Dimensions:						
OD (mm)	250	255	250	205	180	125
Length (mm)	8,160	7,145	7,525	7,550	6,455	2,700
Mass (kg)	2,200	1,800	2,200	1,300	1,000	225

Development trends of HDHMs in the FSU

The information presented in this section is based on a book by Gusman, Lyubimov, G. M. Nikitin, I. V. Sobkina, and V. P. Shumilov, *Calculation, Design, and Operation of Turbodrills* published in 1976. [15] Prof. Gusman, one of the creators of the first multistage turbodrill design was the principal author. The book gives a comprehensive description of the development of HDHM technology in the postwar period and presents some results of its implementation.

Application conditions. In 1949, Russian research engineers resumed work for improving existing turbodrill designs and developing new ones; SKB-2 for developing turbodrill and bit technologies was formed in Moscow. In 1953, special groups were set up within the VNIIBT (formerly SKB-2) structure to develop turbodrills and other types of HDHM designs.

During the postwar period, results of the commercial application of turbodrills in Krasnokamsk (Perm region), Ishimbai in the Bashkiriya region, and in some other regions allowed engineers to identify shortcomings of turbodrills and concentrate efforts to overcome them. In addition, the analysis of these results was instrumental in forming new development trends of HDHM technology.

All improvements of turbodrill designs and new research developments were aimed at meeting the requirements of developing large oilfields. A large number of different types of DHMs were required for various geological conditions to meet specific well drilling requirements.

Turbodrilling was quite important for the development of the main oil-producing regions in the FSU. By implementing the turbodrilling well technology, the Russian oil and gas industry made significant progress in exploration and development of oil and gas fields. Wide utilization of the turbodrilling technology was instrumental in the rapid development of oil producing regions in the Ural-Volga province as well as in discovery and development of new fields in Western Siberia.

Application of the turbodrilling technology while developing oilfields in Tatariya, Bashkiriya, and the Tyumen region resulted in achieving a four to sixfold enhancement of the penetration rate compared to the state-of-the-art rotary drilling rate in the same geological conditions. Meanwhile, the penetration per bit run was 60–100% of the rotary drilling level. As a result, well construction rates increased twofold or threefold at a lower cost per foot.

In the 1950s and early 1960s, turbodrilling technology was used most effectively in the oil and gas fields in the Republic of Tatariya. Local drillers carried out a large number of engineering studies and identified the most efficient drilling practices. High bit weight and rotational speed facilitated efficient rock destruction in the fields of Tatariya. A large amount of field test work was carried out to select a bit type for high-speed turbodrilling applications. Considerable attention was paid to the selection of the correct type of jet nozzle systems to optimize bit hydraulics. The use of water as drilling mud gave new impetus to improvement of bit performance and drilling economics.

In the late 1950s, the Russian drilling industry turned to wide usage of smaller diameter bits for the main well intervals. This was caused by several factors, such as the necessity to drill deep wells with a large number of intermediate casing strings and intensive shallow well drilling in the oil-rich fields. The change was made in order to use 269-mm ($10\frac{5}{8}$-in.), 214-mm ($8\frac{7}{16}$-in.), and 190-mm ($7\frac{1}{2}$-in.) drillbits for the main well interval rather than 295-mm ($11\frac{5}{8}$-in.) bits.

Turbine characteristics and perfection of turbodrill design. Progress in drilling technology required a change in the characteristics of turbodrills and improving their designs. First of all, these design improvements were aimed at achieving a high proportion between torque and rotational speed (M/n). The analysis of the turbodrilling application results allowed engineers to determine these proportions, which set the direction of the turbodrill design development and the turbine flow mechanics research.

Specific operating regimes of small size turbodrill turbines, as well as the multistage turbine pattern, required development of the turbine design theory, as previously noted. The classical studies by Shumilov assisted in getting this work off the ground. His studies, along with development of the issues related to constructing small size multistage turbines, dealt with the range of problems related to fluid flowing through the mud circulation system of the rig. The research engineers used these studies to develop a scientifically based approach to the selection of drilling tools and mud pumps. The specialists from the turbine laboratory of VNIIBT carried out a significant amount of theoretical and test work, which made it possible to calculate reliable turbine parameters and enabled the design of their main power characteristics.

The studies conducted in the FSU and in the West in the mid-1950s indicated the considerable importance of well hydrostatic pressure in deep wells and its strong affect on the drilling process. Therefore, to achieve efficient rock destruction, bit weight had to be increased and rotational speed slowed down to achieve longer bit life.

High torque and low rotational bit speed in the no-gear turbodrill can be achieved primarily by increasing the number of the turbine stages. The first step in this direction was the development of a sectional turbodrill design with a total number of turbine stages of 200 to 500. This took place between 1953 and 1956.

The system that included a turbine with a high degree of circulation (the "slowdown to the brake" trend of the pressure curve) and a valve unit was first suggested by Shumilov in 1936. In 1960, the system was developed and implemented on a practical basis. By reducing the amount of fluid pumped downhole when the differential pressure level in the turbine increased, these types of systems could operate in smooth mud pump drive conditions.

The torque curve inflexion was also achieved in cases of fixed flow rate in certain propeller type wheel turbines, but the gain in torque was accompanied by a rapid differential pressure increase in the turbine.

In 1966, engineers developed a combination turbine design that featured blade rows with both an active profile including a pressure line drop in the slowdown mode and a propeller type profile. These turbines showed the curving torque line at an almost horizontal pressure line.

Other work done by researchers to turn the left part of the high-speed turbine characteristics into a working zone was the development of the turbine hydro-braking system (HBS). By combining a brake vane cascade with a high-speed turbine, engineers managed to build a slow-speed motor with a sufficient torque level. At the same time, the efficiency factor was essentially lower compared to the slow-speed precision cast turbines.

The results of a wide scale testing program of various types of turbodrills indicated that the 3TSSh-195TL spindle-type turbodrills with precision cast turbines met the requirements of the high-speed drilling practices used in oilfields of the Tyumen region. These tests were carried out by specialists from Glavtyumenneftegaz Company. A comparatively low differential pressure in the precision cast turbine allowed the application of jet bits with a differential pressure of 60–80 kg/cm^2, which was quite important for the geological conditions of the Tyumen region. A further increase of the differential pressure level in the jet bit (up to 100–140 bar) during a test at Nizhnevartovsk field led to additional improvements in drilling results. The spindle turbodrill design contributed greatly to successful drilling in the Tyumen region. Taking the poor infrastructure in the region into consideration, the design reduced transportation of the turbodrills to a minimum.

Turbodrill design analysis. The requirement to extend turbodrill service life prompted the introduction of significant improvements in their designs and gave impetus to scientific research work related to rubber-metal and ball bearings.

While working on the issue of turbine hydraulics, research engineers could use the results of a large volume of theoretical and experimental work related to general hydraulic engineering, but they could use nothing except their own experience when studying the issue of bearing friction in an abrasive media. Therefore, they had to develop the main theoretical provisions to calculate turbodrill bearing designs and build special test stands.

Bench tests were carried out for a period of more than 20 years, along with an analysis of the field application results for various types of turbodrill bearings, which enabled development of the scientifically based methods of designing the turbodrill bearings with improved friction characteristics.

The issues related to calculation of the mechanical parameters of the turbodrill design have not been fully resolved yet. The most crucial are issues such as the fit of friction type rotors inside the stator system. Studies by VNIIBT specialists touched upon these issues.

Neither low-pressure turbines nor hydraulic vane cascade brake systems enabled drilling using high torque at rotational speeds of 100–150 rpm. In this regard, the designers were trying to use the volumetric hydrostatic motor design as a basis for developing a low-speed, high-torque motor.

Development of high torque and low-speed hydraulic motors. In 1932, Lyubimov pioneered the development in designing a volumetric DHM. However, neither this motor nor the ones that followed were successful. In 1962, after 12 years of development work, the Smith International Company in the United States built a screw motor called Dyna-Drill. The motor was essentially a reverse single screw pump invented by the French engineer Muano in 1936. Dyna-Drill motors found wide utilization in the United States in directional drilling application. The motor's characteristics differ little from the parameters of the current design turbodrills. Among its advantages are smaller length and lower cost compared to the turbodrills.

In 1966, specialists from the VNIIBT Branch in Perm developed a screw motor design featuring a combination of a multiple thread screw and an eccentric planetary gear. This combination decreased the rotational speed of the rotor, connected through a number of hinges to the spindle shaft, to 100–200 rpm and, at the same time, significantly increased torque. This simple design motor was capable of effectively drilling in combination with certain type bits in rock where high-speed motor drilling proved to be inefficient. One of the interesting features of the new design was the possibility of building a small diameter motor with good power characteristics. For example, an 85-mm motor with unique power characteristics was designed for workover operations and exploratory drilling. This kind of motor, PDM, has wide application nowadays (*see* later sections in this chapter for a description of PDM development).

Along with the development of the PDMs, Russian design engineers carried out considerable work to improve gear type motor designs. Significant success was achieved in improving the system and protecting working elements of the gear from the circulating fluid, which promoted further development for improved geared turbodrill designs. In later sections of this chapter, the current status is presented.

Directional drilling and other turbodrill applications. Growing amounts of directional drilling in the FSU, especially in Siberia, gave impetus to the further development of turbodrilling technology. Furthermore, directional drilling technology is most widely used in offshore drilling operations. Development of new HDHM designs promoted wide utilization of the directional drilling method and assisted in improving its cost efficiency. Both Russian and Western research engineers contributed extensively to the improvement of the directional drilling technology and equipment. They designed special turbodrills for various directional drilling applications. The first chapter of Volume 2 of this series is devoted to this trend of drilling technology.

Especially important is the issue of utilization of DHMs for drilling deep wells. The results of drilling well No. 100 in Azerbaijan in 1965 to a depth of 6500 m, which is described in the following material, proved the significance of this issue. While using turbodrills in combination with diamond bits, drillers achieved a high penetration rate.

Turbodrills of different kinds, including gear reduction, were successfully used for drilling Kola super-deep SD-3 borehole and achieved the world record depth of 12,262 m. The elaboration of successful design solutions for mud pumps and other involved systems continuously exposed to high pressures of 250–300 bar was the necessary prerequisite for improvement of the deep well turbodrilling application.

Based on the turbodrilling application from 1949 to 1958, engineers in the FSU developed the so-called reactive-turbine drilling method for drilling large diameter boreholes for the oil industry as well as for other industries. The new RTB type drills found their successful application especially in the top sections of super-deep boreholes (*see* Volume 2 for more details).

The turbodrilling method in directional and horizontal applications highly promotes its further improvement. Certain challenges must be met, such as drilling extended reach wells (ERW) with targets displaced by several kilometers from the point where the well is spud in. The experience gained in construction to improve the design of DHMs and to develop new drilling technology will assist in the achievement of these goals.

Even if the significant increase in bit life is achieved, cutback or elimination of tripping operations still remains one of the crucial tasks. This becomes especially important while drilling deep or highly deviated directional wells. Development of efficiently working retractable bits run by retrievable DHMs allows the most radical solution for this task. The experience related to the RB technology presented in the last chapter of Volume 2 indicates the possibility of successful applications for this method.

Some of the important trends of the drilling technology enhancement include development of a telemetry system to control DHM parameters and borehole trajectory. The turbodrilling process automation systems should provide for the required bit operating parameters, especially for those related to the bit rotational speed and weight stabilization dependent on the actual well drilling conditions. In addition to the usage of the conventional hydraulic and electrical methods of drilling parameters registration, the research engineers should study the possibility of using acoustic seismic equipment for these purposes.

Initial field tests of Russian turbodrills in the West. The success of the FSU engineers in developing the turbodrilling technology attracted the attention of Western specialists. Using the developments of Russian designers, French engineers built a turbodrill similar to the T14 design.

In 1956, this turbodrill was used in test drilling. Additional research work in the FSU led to the introduction of certain changes to this design. From 1957 to 1959, both French and Russian built turbodrills were used to drill a number of wells in Lak field in France. This project was the first time the turbodrill was used in combination with diamond bits. The bits proved to be quite efficient at high rotational speeds. In 1956, the All-Union Machinoexport Association sold the license for manufacturing turbodrills to Dresser Industries, USA, a drilling industry equipment manufacturer. Two German companies, Salz-Gitter Machinenbau and Ganiel und Lueg, have purchased this license also. All these companies purchased a series of various-sized Russian-made turbodrills.

In 1957, the turbodrill was used to drill oil wells for the first time in Germany. From 1957 to 1959, Dresser Industries carried out work for test drilling in various conditions in the United States using the FSU-made turbodrills. In late 1959, the turbodrills were also tested in Italy.

In most cases, the cost of drilling using single-section and two-section turbodrills with rotational speeds of 600–800 rpm was higher compared to rotary drilling. The Russian-made turbodrills showed good performance results such as long service life of the main components and easy servicing. Nevertheless, these parameters did not affect drilling cost greatly. However, the reduction in penetration per bit run while using turbodrilling resulted in higher drilling costs compared to rotary drilling. Higher penetration rates achieved while using the new drilling method were not sufficient to compensate for the decline in drilled footage. Still, the turbodrilling technology showed better results compared to the rotary method in the following cases:

- drilling through hard, non-abrasive rock
- drilling well sections with limited bit weight to avoid a bottomhole deviation
- directional well drilling
- using diamond bits

Certain factors, such as lack of information about the bit rotational speed and low accuracy of the weight indicator readings, highly complicate the task of establishing required drilling conditions that would assist in achieving better results with the turbodrilling method. Therefore, experience in drilling certain types of wells is of great importance.

The turbodrills exported from the FSU to the Western countries from 1956 to 1959 lacked characteristics that would allow combining high rotational speed with high torque. Therefore, the turbodrilling technology showed successful results only in certain conditions. A similar situation was encountered during the first years when turbodrilling technology was applied in the FSU. Where turbodrilling showed successful results in the Urals and the Volga region, it was significantly less efficient in other regions.

The initial unsuccessful experience with turbodrilling application in the West in no way suggested that the ideas behind the technology were wrong. Still, in most conditions, penetration per bit run had to be increased by 60–70% to achieve good results.

Improvements in the drilling technology and turbodrill characteristics, along with new developments in low speed motor designs, were instrumental in achieving these results. Combining the high rotational speed and high torque, engineers increased the penetration rate fourfold to fivefold. This resulted in the significant increase of the average drilling rate and assisted in lowering drilling costs. Millions of meters drilled in Tatariya and Western Siberia, with average penetration rates of 25–50 m/hr and penetration per bit run of 40–400 m, proved it was possible to achieve these parameters. While drilling in the same conditions in the United States, the penetration rate was 4–10 m/hr.

The application of turbodrilling technology in the West was not limited to the experimental drilling carried out from 1956 to 1959. From 1965 to 1969, a similar project using the Russian-made turbodrills was done in Mexico. In 1965, the three-sectional turbodrill TS5B-7½-in. was used to drill an interval up to 1300 m in depth composed of hard and highly abrasive rock. In these geological conditions, the penetration rate increased fivefold to sixfold, whereas the footage per bit run dropped by 10–20%, which proved to be cost efficient. In 1969, the 3TSSh-195TL with low speed precision cast turbines was used. WOB was low because of the risk of borehole deviation. At a rotational speed of 300 rpm, the penetration rate increased twofold compared to rotary drilling, where the footage per bit run did not change.

Both turbodrills and PDMs were widely used in the West primarily for drilling "build-the-angle" intervals of directional wells. Due to the new generation of drillbits—cone type, PDC, and diamond—that appeared in the last 10 years, different types of PDM and turbodrills became very competitive with rotary drilling for tangent sections as well.

Drillbit selection for turbodrilling. The wide application of bits with sealed bearings set new challenges for turbodrilling technology. The main problem was that the bearing seal assembly of the commonly used bits was not designed to work at high rotational speeds. However, the task here was not to limit the rotational speed of the DHMs in an attempt to adjust it to the operational capabilities of the existing cone bit designs. The task was to develop new, improved sealed-bearing assembly designs to accommodate these types of bits to work with DHMs. The results of pilot applications by the Tatariya and Glavtyumenneftegaz Company using various DHM designs combined with the sealed bearing bits proved the high potential of this development trend.

Drillers from Glavtyumenneftegaz successfully used the 3TSSh-195TL turbodrills with rotational speed of 350–450 rpm in combination with the Russian-made AV type tri-cone bits with bearings capable of operating at high rotational speed and teeth with reduced contact surface. Penetration per bit increased by twofold compared to the best results previously achieved in this region.

On the other hand, the progress made by the artificial diamond industry helped reduce the diamond production costs and promoted their use in manufacturing diamond and PDC bits that are used widely in turbodrilling.

The development of all types of DHMs and bits aim at achieving the main objective, which is the proportional increase of the drilling penetration rate with an increase in rotational speed. Finding technical solutions for this problem involves more than just improving turbodrill and bit designs. It also requires reduction or elimination of differential bottomhole pressure, improvement of bottomhole cleaning, use of optimum volumes of drilling fluid and its flow velocities at the bit jet-nozzle outlets, and improvement of other drilling technology parameters.

The quality of drilling mud plays an important role in this respect. Oil-based drilling mud and LubriFilm drilling mud additives increase the efficiency factor of the motor and bit mechanics. Use of new bearing designs assists in improving oil and heat resistance to enable normal operation in this type of fluid.

More details and the case studies are given in the next sections of this chapter and other chapters in these volumes.

Turbodrill design evolution

This section presents an analysis of the development of the motor designs and types, as well as the ultimate characteristics achieved, thanks to the research work carried out in the 1970s and 1980s. [16]

Turbodrill bearing assembly. Currently, the axial bearing assemblies are among the most crucial elements of the turbodrill designs and determine the success of its application. Quite impressive results, such as the average service life of a bearing assembly, increased several times compared to the previous data and were achieved thanks to successful studies and research for a long period of time. This was possible due to the proper selection of the relevant grade of steel for the axial thrust bearing's discs, heat treatment, surface case-hardening, and the improvement of the rubber lining design of the thrust bearing. In addition, optimizing the number and size of fluid courses to achieve good lubricating effects with drilling mud, enhancement of the used rubber service life, and methods of vulcanization on metal were applied.

In the 1950s and 1960s, the MTBF for the rubber-metal thrust bearing was 10–15 hrs, and for the radial bearing it was about 40–50 hours while drilling with normal density mud. Whereas in mid-1970s, these parameters reached levels corresponding to GOST (State Standard) 4671-70 (*see* Table 2–5), which was five times higher. While drilling using diamond bits, the axial bearing service life increased one and a half to two times thanks to the significant reduction of dynamic vibration loads.

TABLE 2–5
Lifetime of Turbodrill Bearings
Depending on Drill Mud Parameters

Mud Parameters	Mean-time-between-failures (hr)	
	Axial bearing	Radial bearing
Water	100	200
Mud $\gamma < 1.5$ g/cm^3 with sand content up to 2%	65	180
Mud $\gamma < 2.2$ g/cm^3 with sand content up to 3%	55	160

Bear in mind that these results were achieved in the mid-1970s before drillers started using the new technologies that allowed reduction of the drilling mud solid phase to a minimum, meaning the sophisticated mud-processing systems developed later. Currently existing drilling mud technology guarantees a significant increase in this parameter, so 150 hours of life is not a problem for the standard rubber-metal bearing pack even with a mud weight up to 1.7 kg/m^3.

During the 1970s, engineers introduced a number of improvements in the existing turbodrill designs such as the development of thrust-to-seal axial bearings that were placed below the turbine as opposed to the conventional unsealed axial bearings located at the top of the turbine. The new design performed the function of a thrust bearing as well as a seal and could be operated with 50–60 bar differential pressure, which enabled the turbodrill operation to be combined with a jet bit.

Figures 2–19 and 2–20 show the elements of the open-thrust bearing and the thrust-to-seal bearing. Next, new improvements in the rubber-metal thrust-bearing design were introduced. A thrust bearing with a buried rubber lining was designed. This design did away with the rubber that pressed out from under the disc and enveloped its edges and preserved the edges and forms of the fluid circulation courses, which made for better cleaning of the friction surface and better lubricating with the fluid. Figure 2–21 shows the thrust-bearing design with buried (flush-mounted) rubber. The system was successfully used both in the open-thrust bearing and for the thrust-to-seal bearing and resulted in improved parameters compared to the conventional rubber-metal discs.

It is worth noting the successful application of the replaceable rubber elements that proved to be especially efficient in the medium- and lower-radial bearings. This design innovation helped reduce their cost of operation and enhanced their ability to operate in high temperature downhole conditions. Previously, this factor had a negative effect on the bond strength of rubber that was vulcanized to metal,

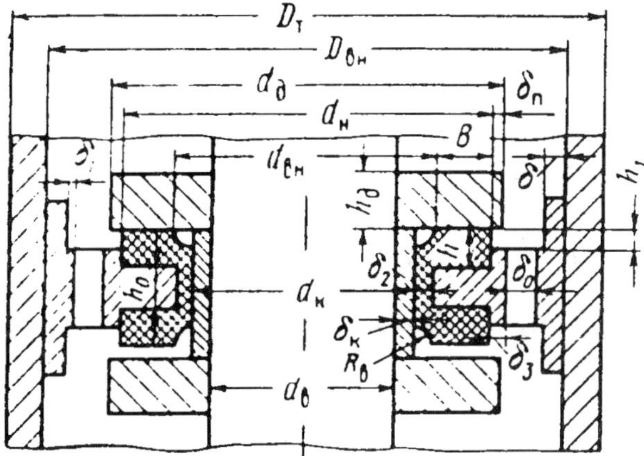

Fig. 2–19 The elements of the rubber-to-metal thrust bearing, "open type"

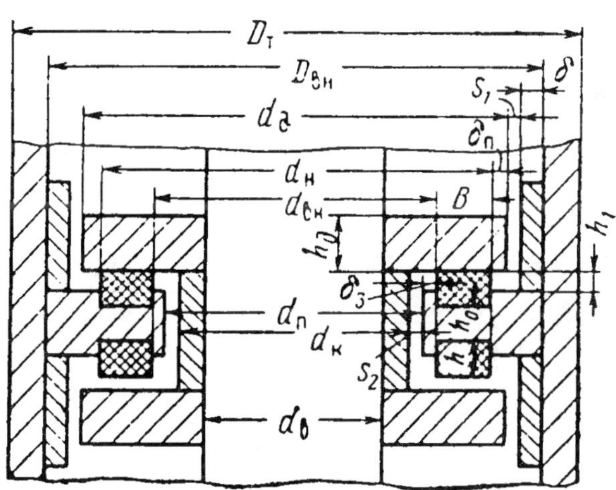

Fig. 2–20 The elements of the rubber-to-metal thrust-to-seal bearing.

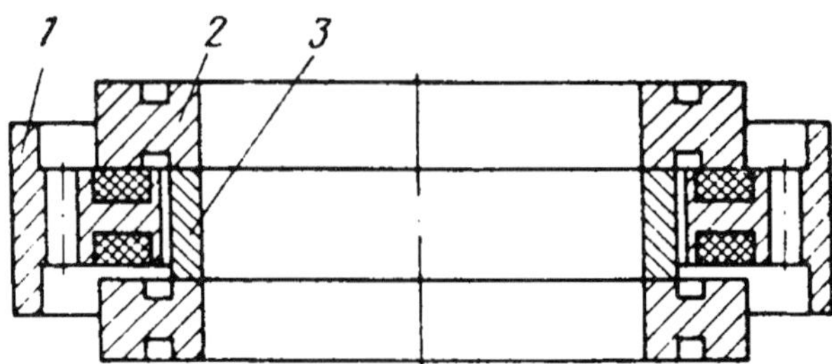

Fig. 2–21 The thrust-bearing design with buried (flush-mounted) rubber

which at 100° C decreases by 40–50% and at 150° C decreases by 60–80% compared to the initial bond strength. When the replaceable rubber elements are fixed mechanically, the temperature limitation for the bonded rubber is eliminated and the use of special heat-resistant rubber material provides the level of service life required of rubber liners at 140–160° C.

The return to the use of frictionless bearings with improved designs—multi-row, non-sealed, lubricated by drill mud—was a serious step toward developing a new solution for the turbodrill thrust bearings. Two types of multi-row (10–20) ball bearings were designed: a double-acting thrust radial bearing shown in Figure 2–22, and an axial double-thrust ball bearing with a rubber-metal bumper shown in Figure 2–23. Both types of the multi-row ball bearings are usually installed on the sectional turbodrill spindle. This type of turbodrill design is analyzed in the following sections.

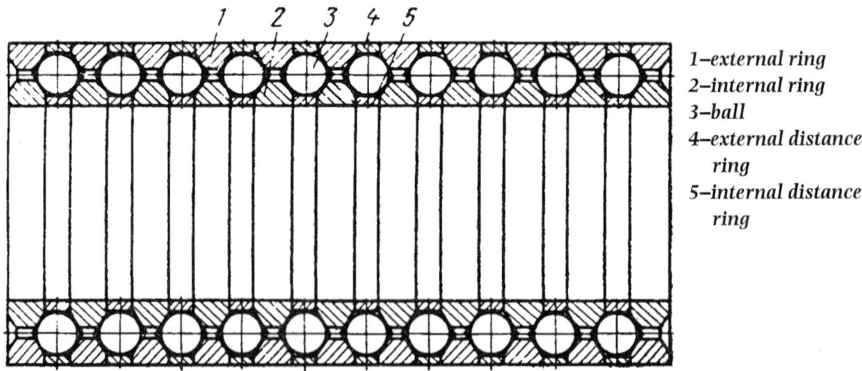

Fig. 2–22 The double-action thrust-radial ball bearing system

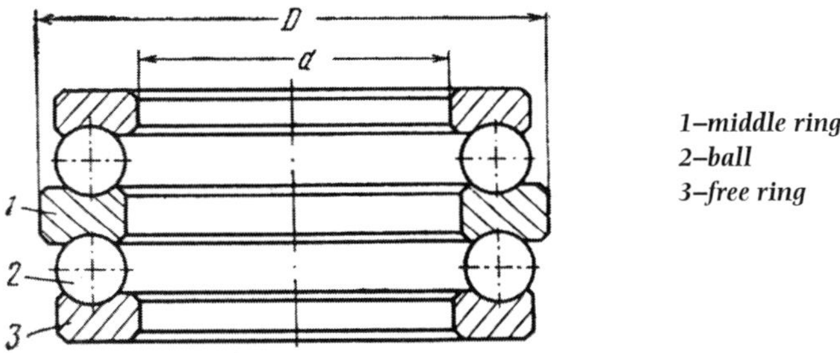

Fig. 2–23 The axial double-thrust ball bearing with a rubber-metal damper

When drilling with roller-cone bits, the ShShO-type design with a spindle, double-thrust ball bearings, and rubber bumpers (shock-absorbers) provided a more uniform load distribution between the bearing rows and resulted in an increased bearing life compared to the conventional rubber-metal thrust bearing design (Table 2–6).

TABLE 2–6
Comparative Test Results for ShSh01-195, ShSh01-172, and Sh1 Spindles (Comparison made by the Spindles Operated with Identical Types of Drill-bits)

Company	Spindle	Interval (m)	Total number of bearings in operation	Drilling volume (m)	Drilling time (hr)	Number	Drilling volume (m)	Drilling time (hr)	MTBF (hr)
Nizhne-volzhskneft	ShSh01-195	800-3,200	22	14,164	4,270	5	3,220	970	194
	Sh1-195	800-3,200				39	6,320	2,180	55
	ShSh01-172	1,770-3,200				4	1,810	824	206
	Sh1-172	1,770-3,200	8	2,324	2,324	21	1,385	657	31.3
Tatneft	ShSh01-195	300-1,700	10	13,200	836	3	6,697	470	156
	Sh1-195								
	ShSh01-195	300-1,700				42	43,000	3,440	82
	Sh1-195	300-1,700	8	15,743	1,375	2	4,260	434	217
Bashneft	ShSh01-172	100-2,150	7	13,379	1,283	2	5,713	450	225
	Sh1-172	100-2,150				48	58,600	5,100	104
Permneft	ShSh01-172	250-2,100	2	4,988	730	2	4,988	760	380
	Sh1-172	250-2,100				15	4,720	705	47

The evolution of the thrust-bearing designs contributed greatly to the development of new turbodrill designs used through the 1990s. Some of their designs and characteristics are presented as follows.

Single-section turbodrills. The T14-9¾-in. turbodrill was the main type turbodrill used during World War II and the postwar periods. As mentioned before, this design had the advantage of using a rubber-metal collar thrust bearing that was separated from the turbine and installed in the upper part of the turbodrill. The thrust bearings were fixed in place by the upper sub and joined to the turbodrill housing through the straight thread, which was the weak link of this design. Because of vibration, the threads usually twisted off, and the turbodrills were left in the hole.

The drillers had to tolerate this while bearing service life was low. It required frequent inspection and replacement of the bearing at the rig. However, when the rubber-metal bearing service life achieved 50–100 hours, the designers had an opportunity to change the design to eliminate the drawback. They returned to the T12 turbodrill design, relocated the thrust bearing inside the turbodrill housing, and fixed it to the turbine using a nipple. Thus the T12M1, T12M2, and T12M3 turbodrills appeared following a series of modifications and tests. The latter model became the main one in batch production of single-stage turbodrills. This design made it possible to connect the crossover sub to the turbodrill housing by means of a high-powered, thick-walled tapered thread, which simplified the turbodrill design considerably and excluded the risk of failure. Figure 2–24 shows the T12M3B turbodrill, the characteristics of which are presented in Table 2–7.

TABLE 2–7
T12M3B Turbodrill Characteristics

Type of TD	Stage number per turbine	Mud flow (l/sec) γ=1g/cm³	Shaft Rotation speed (rpm)		Torque (kg.m)		Power (HP)	Pressure drop at maximal power (kg/m²)
			At max. power	Free run	At max. power	Stall		
T12M3B-240	104	50	660	1,320	200	400	185	40
		55	725	1,450	240	480	240	45
T12M3B-215	99	40	545	1,090	110	220	85	25
		45	610	1,220	140	280	120	35
T12M3B-195	100	30	660	1,320	85	170	80	35
		35	770	1,540	115	230	125	45
T12M3E-172	121	25	625	1,250	65	130	55	30
		28	700	1,400	80	160	75	40

In the 1950s, when the amount of turbodrilling and its proportion in the total drilled footage was rapidly growing, certain major drawbacks of the application of

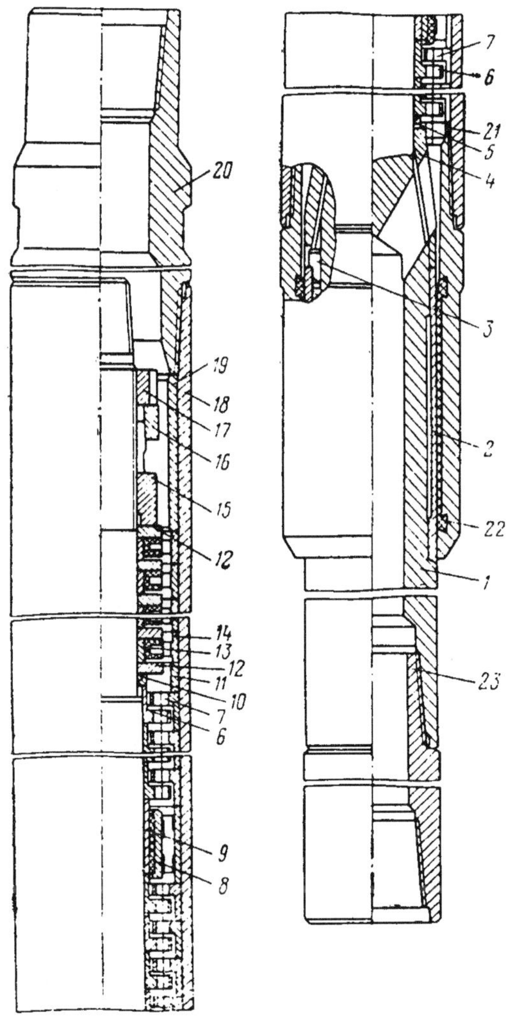

Fig. 2–24 The T12M3B turbodrill

a multistage turbine were identified. The system required continuous fine adjustment of clearances between the rotor and stator discs, which were hard to accommodate because the axial bearings wore out. In addition, continued tightening of the stator reduced, resulting in stator twist and slump. These negative factors led to axial wear of the turbine discs, which in turn necessitated removal of the turbine from the housing to replace the discs.

Further, this operation frequently was complicated because in the course of time, the drilling mud got in the clearance between the stator rims and the turbodrill housing and formed a firm cement crust, which then required a great deal of force to remove the turbine. Given the shortage of service and repair bases for turbodrills that existed at that time, they had to be repaired at the rig site using a drawworks that was capable of providing the required force. When the force created by the drawworks was not sufficient, drillers utilized a special method, shown in Figure 2–25 [17], that employed a DS weight. At the repair depot, failed turbines were removed using a hydraulic press.

These difficulties prompted another design idea—development of a multistage turbine with a radial-axial fluid inlet to the vane system instead of an axial inlet.

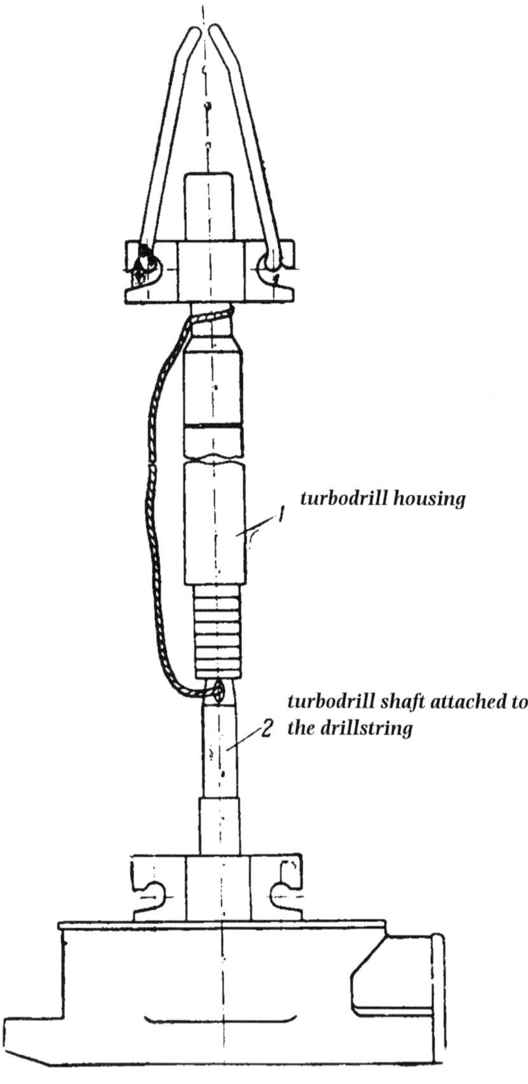

Fig. 2–25 The method of turbodrill disassembling at the rig floor

This allowed construction of a no-end-face no-crown vane cascade turbine with stator and rotor blades located at various diameters (Fig. 2–26[18]). This design afforded free movement of the shaft with the turbine rotors inside the housing with built-in stators. Displacement of the rotor mean position by a maximum of 3.5 mm from the stator did not cause any significant change in the turbine characteristics, but the efficiency of the radial-axial turbine was not higher than 50% because of the change in flow direction at each stage.

The T25-9¾-in. turbodrill (Fig. 2–27[19]) was commercially manufactured and mostly used in Tatariya, Bashkiriya, and in the Samara region, where they proved to be quite efficient. For example, the MTBF for this type turbine in some cases reached 500–600 hours. Its maintenance was quite simple and resulted in the decrease of their operative cost by 15–20% compared to turbodrills with the conventional axial flow turbine. However, the analysis of their operation results indicated that their efficiency was reduced during the course of time, which was caused by a larger radial clearance between the rotor and the stator. The efficiency

Fig. 2–26 Turbine stator and rotor of radial-axial turbodrill

factor was reduced as much as 35–40%. Because of the greater length of the stage in this type turbodrill, the total number of stages was lower, which negatively affected torque. The negative moments outweighed the positive ones, which prompted drillers to refuse to use this type of turbodrill for regular applications.

Yet, turbodrills were successfully applied in casing drilling using retractable bits where they proved to have significant advantages compared to regular turbodrills (*see* the last chapter in Volume 2 for details).

Single-section turbodrills were subjected to a new stage of development after the introduction of turbodrills with the spindle type design and the new type of turbine (TVSh-type) in the late 1980s and 1990s. These are described later in this chapter.

Sectional turbodrills. The growing amount of deep well drilling (wells of 3000–3500 m) again brought up the issue of increasing torque and reducing the turbodrill shaft rotational speed. The theory of multistage turbodrills indicated that the increase in the number of stages was the main option for achieving this goal while using turbodrills without gear reduction.

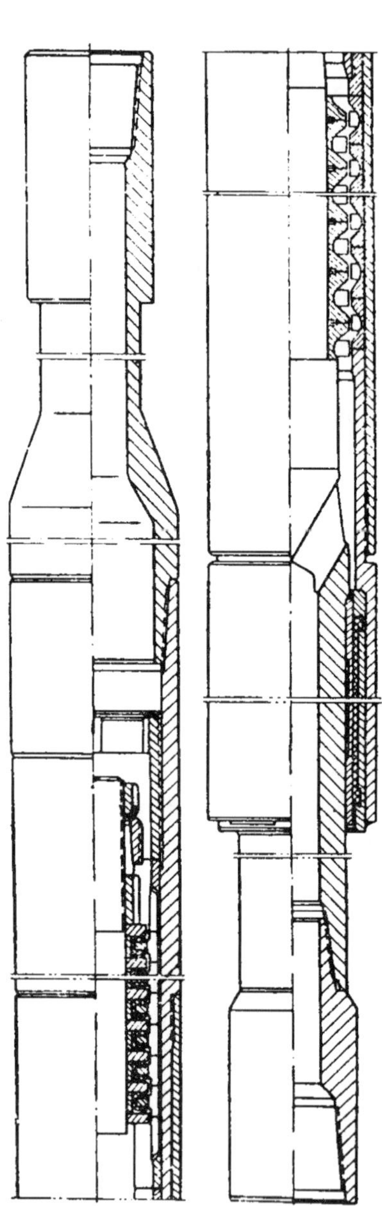

Fig. 2–27 The T25 $9^{3}/_{4}$" Turbodrill with radial-axial turbine

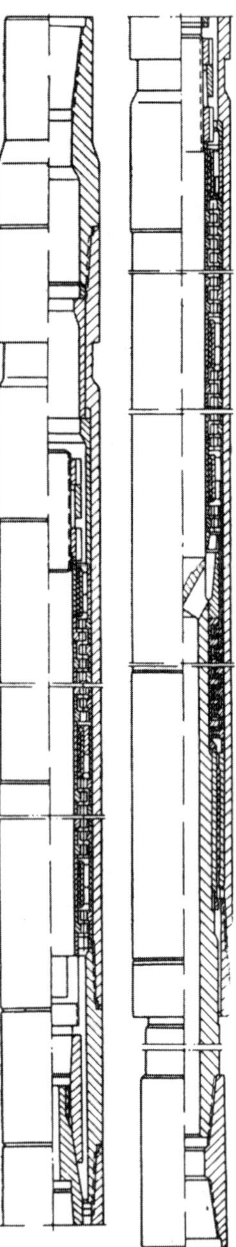

Fig. 2–28 The TS6 ($6^{5}/_{8}$") two-section turbodrill

In the case of the single-section turbodrill, this option faced a number of essential problems related to design features as well as inconveniences associated with the manufacture and operation of long turbodrills (more than 10–11 m). For this reason, the engineers conceived of a sectional turbodrill during 1940 and 1941 that was based on the theoretical works by Shumilov. However, the practical implementation of this idea did not commence until the late 1940s and early 1950s.

During this period, the research and design engineers developed and built first a two-section turbodrill, followed by the three- and four-section turbodrills. The TS1-8-in. turbodrill, which was the first two-section turbodrill capable of operating, was field-tested in 1953 near Baku. The tests revealed a number of drawbacks in the components design, such as the shafts joint assembly, bearings and their location, etc.

From 1954 to 1955, the engineers developed and built the TS3 and TS4 turbodrills in an attempt to eliminate the drawbacks. These turbodrills featured 180–200 stages. In 1956, drillers commenced using the first three-section TS4-5-in. turbodrill with 240 stages. Next, the three-section 3TS5B and 3TS5E turbodrill with diameters of $6\frac{5}{8}$-in., $7\frac{1}{2}$-in., 8-in., and 9-in. were successfully tested.

An important design solution was implemented during this period that concerned the relocation of the axial bearing to the lower part of the turbodrill. This was built as a no-flow-through thrust bearing, which made it possible to use jet bits (*see* discussion later in this chapter), and was realized in the TS6 ($6\frac{5}{8}$-in.) two-section turbodrill design shown in Figure 2–28 [20].

The field and bench tests of this turbodrill design indicated that at a differential pressure level of 40–50 bar in the bit, fluid leakage did not exceed 2–3% compared to 40% in turbodrills with a conventional radial bearing. Table 2–8 summarizes the characteristics of the TS-type sectional turbodrills that feature the lower axial no-flow rubber-metal thrust bearing, including TS4A-127 and TS4A-104.5 turbodrills, used to drill small diameter exploration wells and for workover operations.

TABLE 2–8
Characteristics of the Two- and Three-Section Turbodrills

Type of TD	Stage number per turbine	Mud flow (l/sec) γ=1g/cm³	Shaft Rotation speed (rpm)		Torque (kg.m)		Power (HP)	Pressure drop at maximal power (kg/m²)
			At max. power	Free run	At Max. power	Stall		
TS 5B-240	210	38	500	1,000	230	460	160	45
		40	525	1,050	260	520	190	50
TS 5B-215	212	30	405	810	130	260	75	30
		35	470	940	175	350	115	45
TS 5B-195	177	25	550	1,100	100	200	78	40
		28	615	1,230	130	260	110	50
TS 5E-172	239	20	500	1,000	80	160	55	40
		22	550	1,100	95	190	70	50
3TS 5B-240	311	32	420	840	250	500	150	50
		34	450	900	280	560	180	55
3TS 5B-215	325	28	380	760	175	350	95	45
		30	405	810	200	400	115	50
3TS 5B-195	272	22	485	970	120	240	80	50
		24	530	1,060	145	290	105	60
3TS 5E-172	352	18	450	900	90	180	60	50
		20	500	1,000	115	230	80	60
TS4A-127	240	12	740	1,480	35	70	35	50
		13	800	1,600	40	80	40	60
TS4A-104,5	212	8	870	1,740	15	30	20	45
		9	980	1,960	20	40	25	55

Placement of the thrust bearing at the lower part of the turbodrill shaft had two faults:

1. the reduced diameter of the turbodrill shaft at the bearing installation area (and consequently the shaft's strength)
2. operational inconveniences associated with the need to remove the turbodrill shaft when replacing the bearing

These shortcomings were eliminated in the 3TSSh-type turbodrill design that incorporated a number of important achievements in the turbodrill design process (Fig. 2–29). The spindle design, preserving all the advantages of the TS6, allowed significant simplification of manufacturing and operation of the turbodrills. By moving an axial bearing into an independent spindle assembly, the design engineers facilitated and sped up the assembly replacement, which could be done now directly at the rig site without disassembling the turbine section. This system significantly reduced transportation costs associated with hauling turbodrills to repair bases.

Furthermore, it relieved these bases from performing preventive maintenance on the motors and the need to repair housings, shafts, and other components. This allowed them to focus on the major turbodrill repairs that were required after prolonged operation. The spindle-type turbodrill application proved their efficiency and promoted their further wide utilization.

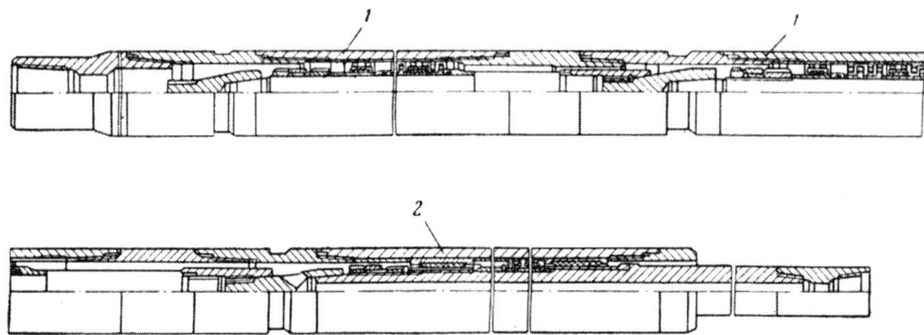

Fig. 2–29 Turbodrill 3TSSh-type

A significant step forward in the enhancement of turbodrill designs was the improvement in manufacturing quality of turbine discs—especially blade quality. The methods used initially, one-piece cast steel discs using sand molds, were very labor-intensive and failed to produce high quality blades, both in terms of the outer edge thickness (1–1.2 mm) and the surface cleanliness that determined the hydraulic loss level.

In addition, this method of sand mold casting did not ensure that the cast turbines met the parameters of the prototype. Therefore, characteristics of the batch production turbodrills were often significantly different from the requirements. A change from the sand mold casting to the precision cast method (melted model casting or wax casting) brought about considerable improvement in the quality of the blade system. The turbine batch production method differed from the technique used originally because the precision cast method sometimes, especially for large size turbines, was used only for production of the circular flow-through element (blades crown). Hubs were made of round billets and then were jointed to the setting section using a hot-pressing technique. The studies performed indicated that the hydraulic friction factor of the precision cast turbines was 0.15–0.18 compared to 0.26–0.33 for the conventionally manufactured turbines.

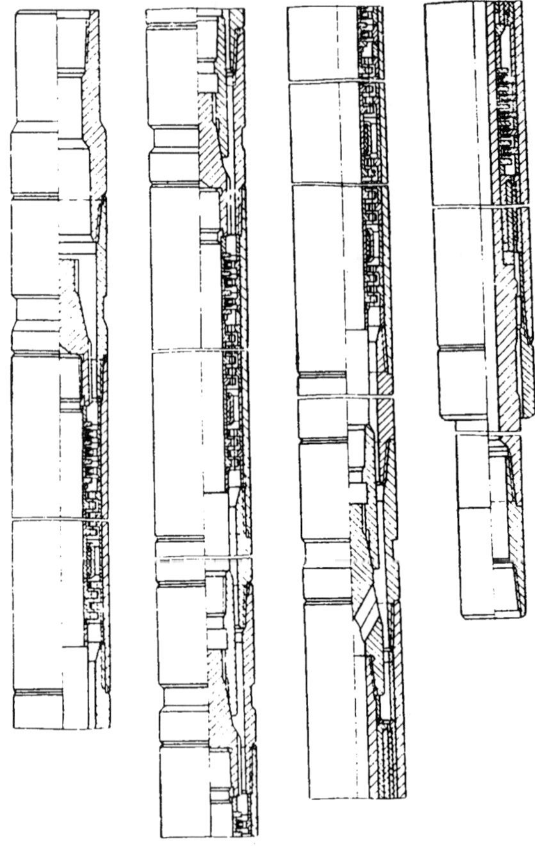

Fig. 2–30 Unified turbodrill 3TSSh-1 with a spindle that has a rubber-metal thrust bearing

Blade crowns made of polymer materials proved to be quite efficient. The crown itself was essentially a replaceable part installed in reusable steel hubs. The polymer turbine blade system featured extreme cleanliness of the flow surfaces and precision in the geometry and size of the blades. These qualities ensured a low hydraulic friction factor of the turbine stages. The characteristics of the batch-produced turbine and their pilot prototypes were almost the same, which added the desired stability characteristics. A major advantage of plastic turbine was its low cost compared to steel stages and its simple and easy manufacturing technology. Polymer utilization required a large volume of research work aimed at developing a material that met the tough requirements of the drilling technology, such as high strength and resistance to vibration, heat, wear, and chemicals. The material composed of Polyamid-12 met these requirements, which assured the successful commercial application of turbines made from it.

Further improvement of the three-section turbodrill design resulted in development of the unified turbodrill 3TSSh-1. Depending on the actual drilling conditions, that machine provided an opportunity to use all the turbodrill design innovations developed by that time that had been tested and used in commercially manufactured turbodrills. Figure 2–30 presents the design of this turbodrill with rubber-metal thrust bearing. Figure 2–31 shows the spindle-only with two types of bearings: radial-thrust ball bearing and thrust ball bearing with rubber shock absorbers.

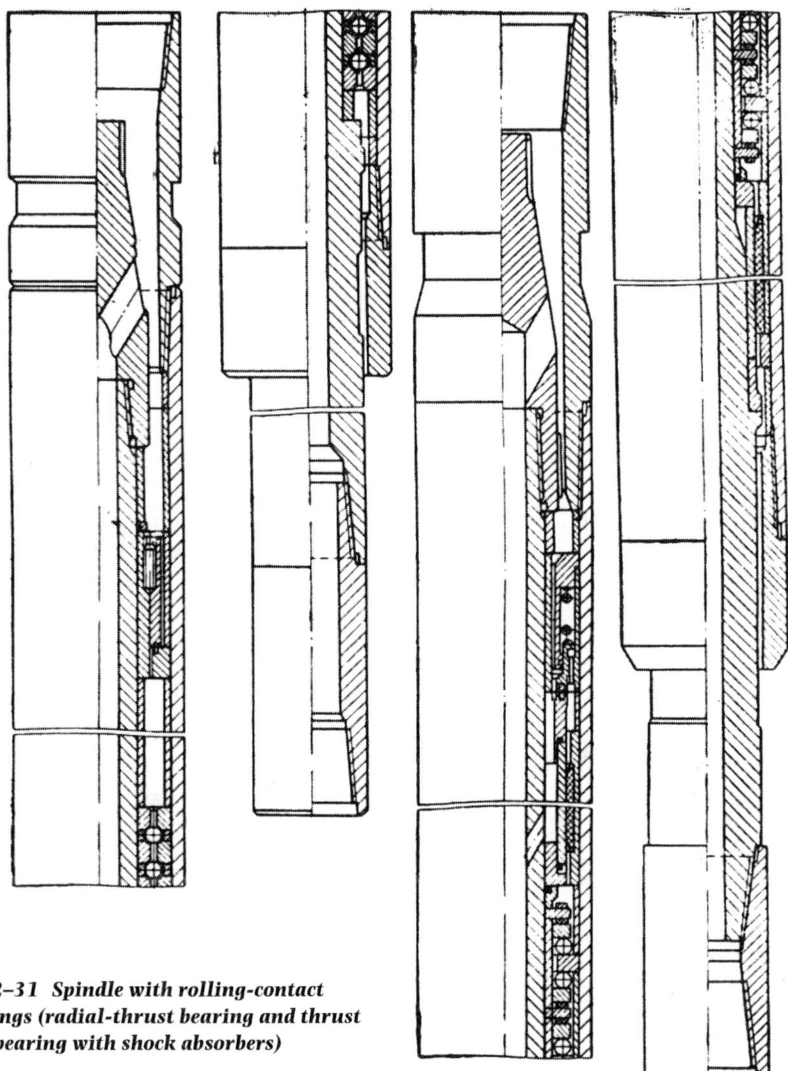

Fig. 2–31 Spindle with rolling-contact bearings (radial-thrust bearing and thrust ball bearing with shock absorbers)

The characteristics of the sectional spindle turbodrills are given in Table 2–9. By the mid-1970s and in the 1980s, these types of turbodrills were used most widely by drilling companies in the FSU—especially in Western Siberia. The data in Table 2–10 indicates that utilization of three-section turbodrills compared to single-section turbodrills, allowed for a 1.4–1.6 times reduction of flaw rate. Similar reduction rates were achieved in the shaft rotational speed, at the same time preserving about the same power because of the significant increase in torque. In these conditions, the differential pressure in the sectional turbodrill increased slightly because the number of stages was approximately three times higher, whereas the differential pressure level in each stage decreased about 2.25–2.5 times. This differential pressure increase in the turbodrill is amply compensated by the general reduction of hydraulic losses in the entire circulating system as a result of the flaw rate decrease.

TABLE 2–9
Characteristics of the Section Spindle-type turbodrills

Type of TD	Stage number per turbine	Mud flow (l/sec) γ=1g/cm^3	Shaft Rotation speed (rpm)		Torque (kg.m)		Power (HP)	Pressure drop at maximal power (kg/m^2)
			At max. power	Free run	At Max. power	Stall		
3TSSh-240	318	32	420	840	250	500	150	50
3TSSh1-240		34	450	900	285	570	180	55
3TSSh1-240TL	318	40	195	390	170	340	50	16
		45	220	440	210	420	65	20
3TSSh-215	333	28	380	760	180	360	100	45
		30	405	810	205	410	120	50
3TSSh-195	306	30	400	800	130	260	75	35
3TSSh1-195		35	470	940	180	360	115	40
3TSSh-195TL	318	40	355	710	175	350	85	30
3TSSh1-195TL		45	400	800	220	440	120	40
3TSSh-172	336	20	505	1,010	100	200	70	60
3TSSh1-172		22	555	1,110	120	240	93	73
3TSSh-164TL	348	23	440	880	110	220	70	50
		25	480	960	130	260	85	55

TABLE 2–10
Characteristics of One- and Three-sectional Spindle Turbodrills Comparison

Type of TD	Stage number per turbine	Mud flow (l/sec) γ=1g/cm³	Shaft Rotation speed (rpm)		Torque (kg.m)		Power (HP)	Differential pressure at max. power (kg/m²)
			At max. power	Free run	At Max. power	Stall		
T12M3B-240	104	50	660	1,320	200	400	185	40
		55	725	1,450	240	480	240	45
3TSSh-240	318	32	420	840	250	500	150	50
3TSSh1-240		34	450	900	285	570	180	55
3TSSh1-240TL	318	40	195	390	170	340	50	16
		45	220	440	210	420	65	20
T12M3B-215	99	40	545	1,090	110	220	85	25
		45	610	1,220	140	280	120	35
3TSSh-215	333	28	380	760	180	360	100	45
		30	405	810	205	410	120	50
T12M3B-195	100	30	660	1,320	85	170	80	35
		35	770	1,540	115	230	125	45
3TSSh1-195	306	22	470	940	180	360	115	40
3TSSh1-195TL	318	45	400	800	220	440	120	40
T12P3E-172	121	25	625	1,250	65	130	55	30
		28	760	1,400	80	160	75	40
3TSSh1-172	336	22	555	1,110	120	240	93	73
3TSSh1-164TL	348	25	480	960	130	260	85	55

Deep drilling application. Significantly lower rotational speeds of the turbodrill shaft and bit, along with increased torque, promoted use of three-section turbodrills in deep wells. The first experience using three-section turbodrills for drilling at depths of 4500–5000 m in complicated geological conditions were well Nos. 144 and 153 of the Karadag field in Baku, which proved to be quite positive. [21] The 3TS5B-9-in. turbodrills were used for drilling well No. 144 in 1961 and 1962. Turbodrilling was applied in the interval below the 11-in. intermediate casing string from 2200 to 4800 m TD. Table 2–11 presents a comparison between the performance results from this well and the earlier best rotary well under the same conditions.

TABLE 2–11
Turbodrilling in Well No. 144 (Karadag Field, Azerbaijan) Comparison with the Earlier Best Rotary Drilling Results in the Same Conditions (1961–1962)

Borehole No.	Drilling interval (m)	Number of runs	Footage per bit (m)	ROP (m/hr)	Drilling run rate (m/hr)
144	50-4000	174	24.9	4.20	1.63
166	110-3898	99	34.5	2.42	1.45

After these results, the 3TS5B-9-in. turbodrill was planned for use in drilling test well No. 153 in the same field. The drilling program provided for utilization of every technique that would allow the most efficient application of turbodrilling. Among them were jet bits with various types of cutting structures, special bottomhole assemblies (BHAs) to increase bit weight, and mud pumps with higher pressure capacity. Drilling this well was planned within the framework of a program to drill three test wells each using a different method: turbodrilling, rotary drilling, and electrodrilling (*see* Chapter 3).

The interval below the 11-in. casing of this well from a depth of 2300 m was drilled mostly using the 3TS5B-9-in. and partially a prototype of the 3TS7-8-in. turbodrill. In some intervals, the actual bit rotational speed was measured using a pulse tachometer. For example, in the interval from 3654 to 3664 m, while using the 3TS5B-9-in. turbodrill with a ball-type spindle, at 16–18 tons bit weight and 36 l/sec flow rate, the registered bit rotational speed was 435 rpm. While drilling with the 3TS7-8-in. turbodrill in the interval from 4091 to 4107 m, the rotational speed was 350–355 rpm, which corresponds to the parameters shown in Table 2–8.

While drilling the interval below the 8-in. liner from a depth of 4534 m and using 1.85 g/cm^3 density mud, the 3TS6-6$\frac{1}{2}$-in. turbodrill was used in combination with cone bits and, in the lower part of the interval, with diamond bits. Table 2–12 presents the comparison between the drilling results from well No. 153 and the two adjacent wells that were drilled using the rotary method.

TABLE 2–12
The Comparison Between the Turbodrilling Results from Well No. 153 (Karadag Field) and the Two Adjacent Wells Drilled Using the Rotary Method

Borehole No.	153	199	187
Borehole depth (m)	4,745	4,853	4,647
ROP (m/hr)	3.0	1.41	1.75
Drilling run rate (m/hr)	1.01	0.58	0.75
Number of runs	266	396	309
Footage per bit below 3000 m depth (m)	14.0	6.2	7.9

Note: Drilling run rate takes into consideration tripping time - vR = run length / (drilling time + tripping time).

After the drilling of well Nos. 144 and No.153 was finished, hundreds of wells were drilled across the entire Karadag field (except for the earlier developed part), using three-section turbodrills. This was the beginning of the wide scale utilization of turbodrilling technology for drilling deep wells in all regions of the FSU. This technology was used most intensively for offshore vertical and directional applications in the Azerbaijan sector of the Caspian Sea.

The results of lower interval drilling in well No.153 confirmed the problems encountered while using the turbodrilling technology with drilling mud density that exceeded 1.5–1.6 g/cm^3. The problems encountered were specially studied by Drilling Department No.3 of the Turkmenburneft Company in 1953 while drilling well No. 455 in Turkmeniya. [22] Barite ($BaSO_4$), a mineral with a high specific gravity of 4.3–4.5 g/cm^3, was used as a drilling mud-weighting additive. The pulverized barite did not contain large and abrasive particles.

The T12M1-8-in. turbodrill was used to drill this well. The drilling revealed that as a rule, the turbine was plugged up and the pressure increased because of continuous interruptions of the drilling process and mud circulation, as well as during the procedure to increase mud weight. During such breaks in the drilling process, barite particles precipitated from the mud, built up on the blades, and plugged up the turbine. Therefore, the number of interruptions to mud circulation while drilling should be reduced to a minimum, and mud should be thoroughly cleaned from the solid phase. Also, mud should be circulated more frequently through the well and turbine during tripping operations. In the case of a forced interruption of the drilling process, prior to breaking circulation, the DS should be rotated with the turbodrill housing to stall the turbodrill in conditions of significant weight simultaneously applied to the drillbit. This would result in the rotation of stator discs around the fixed rotor discs.

By measuring the turbodrill shaft speed during the test at various pump rates, it was determined that friction of the axial rubber-metal bearing caused a significant loss of power. Using turbodrills with ball bearings for drilling with heavy mud is recommended.

In the case of well No. 455, the drilling mud density was maintained at 1.8 g/cm^3. Extensive experience in deep well turbodrilling allowed drilling wells with a mud density of 2.0 g/cm^3 and greater. Yet the current official recommendations limit the mud density level to 1.7–1.8 g/cm^3.

Experience with precision-casting turbines in Western Siberia. As mentioned, the precision-cast turbines proved to be very successful while drilling in Western Siberia. The precision-cast low-speed turbine has a number of design features related to the need to reduce the rotational speed. Among these features are the high blade pitch angles of 72°–75° compared to 62°–65° in the regular turbines. Another feature is the thin outer rim of the blade.

Data in Table 2–9 indicates that because of these factors and the smoother blade surface, the differential pressure went down Rather than up for turbodrills with precision cast turbines at higher flow rates of 30% or more. In addition, the rotational speed of the turbodrill shaft essentially decreased, whereas torque went up. The possibility for increasing the circulation rate was quite important for drilling wells in Western Siberia because it allowed directional well cleaning improvement, especially in wells with a high degree of inclination. Also, a pressure decrease in the turbine enabled utilization of jet bits.

Figure 2–32 [23] shows an example of two different wax casting turbine profiles. The design featured a shorter stage and blade length and allowed for a significant increase in the number of stages per turbodrill section, which allowed significant changes in the turbodrill characteristics as presented in Table 2–13.

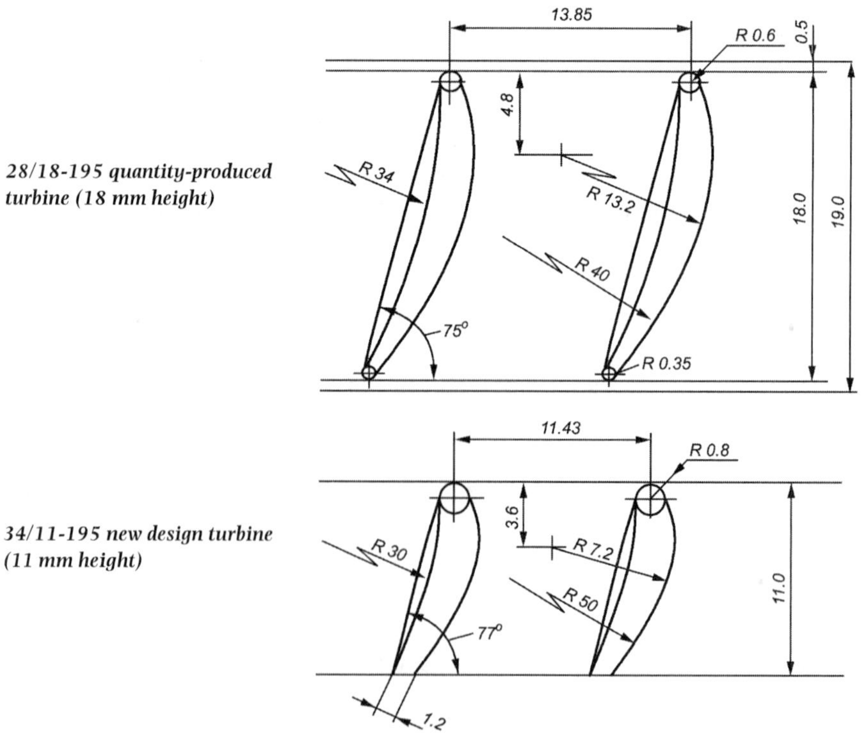

Fig. 2–32 Precision-casting turbine stages of different profiles

TABLE 2–13
Comparison of 3TSSh-195TL Turbodrill with the Turbines 28/18-195 and 34/11-195

Type of TD	Turbine type blades number/ blade thickness)	Number of sections	Number of stages per turbodrill	RPM	Torque (kg m)	Pressure drop kg/cm²)	Effective power (HP)	Efficiency factor
3TSSh-195TL	28/18-195	3	330	259	117.5	18.7	41.1	0.55
3TSSh-195TL	34/11-195	3	445	290	153.5	26.9	51.4	0.48

Turbodrills featuring pressure descending towards the stall. As mentioned in this chapter, turbines with a degree of circulation $\sigma > 1$ still would find an application because the differential pressure in the turbine at reduced rotational speed allowed the pumping of more fluid through the turbine at low rotational speed and resulted in higher torque. Drillers can control the mud circulation rate through the turbine, characterized by the pressure curve descending toward the braking mode (stall), using special downhole or surface devices. Flexible drive mud pumps, which are direct current (DC) electrical motors and a diesel engine with a transformer turbine, control the circulation rate at the surface. Special equipment such as pressure-reducing valves, ejector type hydraulic flow multipliers, or flow dividing systems enables downhole mud circulation control.

At first, turbodrills with such turbines were equipped with pressure-reducing valves. In large diameter turbodrills, the valve was installed in the turbodrill hollow shaft, whereas in small diameter turbodrills, it was installed in the specially attached section above the turbodrill.

Turbodrills characterized by a declining pressure curve later became widely used at constant flow rates without the pressure reducing valves even though the characteristics of these turbodrills became slightly worse. The pump outlet pressure varied depending upon the operating regime.

The industry produced three types of turbodrills with declining pressure curve characteristics—A9K5Sa, A7N4S, and A6K3S with ODs of 240 mm, 195 mm, and 164 mm respectively.

The A9K5Sa and A7N4S turbodrills (Fig. 2–33) were built similar to the TS6 sectional turbodrills but differed from the latter in bearing assembly design. A rubber-metal bearing was replaced with the multi-row (12 to 15 rows) axial thrust ball bearing (*see* Fig. 2–22) that took up all axial loads and was installed in the turbodrill lower section. Single row radial ball bearings were installed in the midsection of the turbodrill. The end seal, installed above the radial-thrust bearing, prevented large abrasive particles from getting inside the bearing by limiting the mudflow rate through it. The seal allowed for utilization of turbodrills in combination with high-pressure bits.

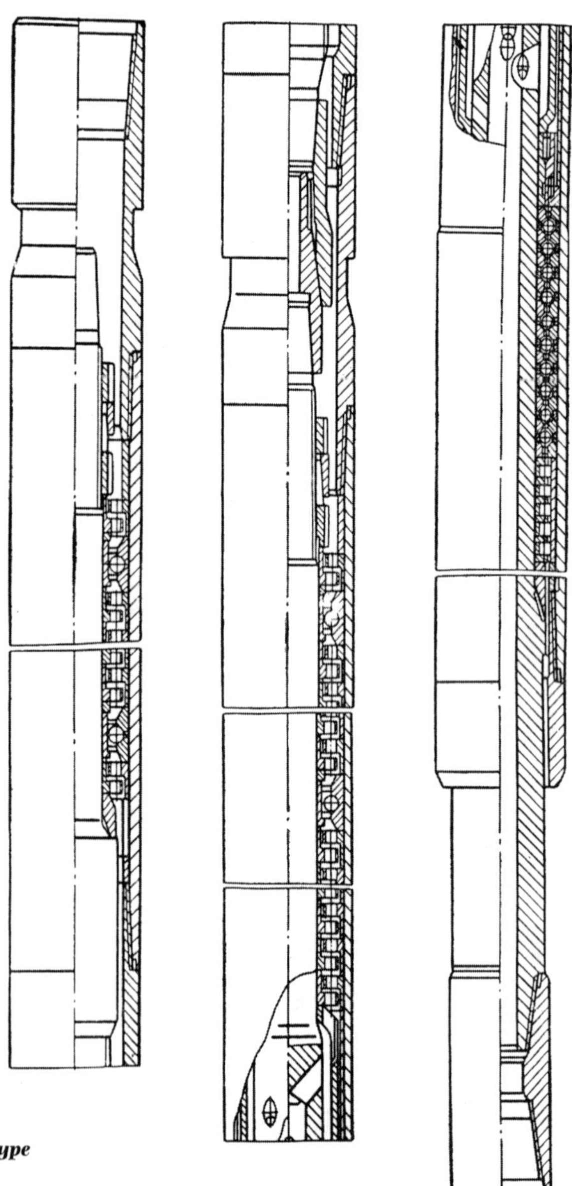

Fig. 2–33 Turbodrills of "A" type

The A6K3S turbodrill design featured an independent shaft suspension on axial bearing in each section. The 10-row radial-thrust ball bearing took up the hydraulic load in the upper section. The radial-thrust bearing in the lower section took up the hydraulic load and bit weight.

The HBS was used to reduce the shaft rotational speed. The system consisted of stators and rotors with blades that (unlike the turbine stator and rotor blades) had an angle identical to the plane and perpendicular to the turbine shaft axis. In the HBS, part of the hydraulic flow energy transformed into mechanical energy that was used for the turbodrill shaft braking. The HBS installed in the turbodrill took off a certain amount of torque created by the turbine. The higher the turbine rotational speed, the more torque was taken off by the HBS stages. By varying the proportion between the number of turbine stages and the number of HBS stages, it was possible to reduce the shaft rotational speed in the no-load mode to the required level. However, a reduction of the shaft rotational speed in the speedup mode occurred due to a power decrease in the turbodrill and resulted in a lower efficiency factor. Three types of turbodrills equipped with the HBS were produced commercially: A9GT, A7GT, and A6GT (with corresponding diameters of 240 mm, 195 mm, and 164 mm).

The A9Sh, A7Sh, A6Sh, A9GTSh, A7GTSh, and A6GTSh turbodrills, equipped with an HBS, were manufactured with the axial bearing fit in a separate spindle similar to the 3TSSh sectional turbodrills previously described. Table 2–14 shows characteristics of the A-type turbodrills operated using water circulation.

TABLE 2–14
Turbodrills with Descending Pressure Curve Characteristic

Type of TD	Number of turbine/ hydro-breaks per turbine	Mud flow (l/sec) γ=1g/cm^3	RPM		Torque (kg m)		Max. Power (HP)	Pressure drop at max. power (kg/m^2)
			At max. power	Free run	At max. power	Stall		
A9K5Sa	203	45	200–30	600	200–300	610	90	50
A7N4S	226	33	300–500	1,000	180–280	455	130	70
A6K3S	220	18	300–400	1,200	70–90	150	34	70
A9GT	334/104	45	250	620	345	690	122	64
A7GT	353/119	30	300	725	190	380	80	72
A6GT	335/121	20	280	620	75	150	28	41
A9Sh	210	45	400	975	310	620	180	68
A7Sh	236	30	520	1,200	190	380	140	82

Turbodrills for diamond drilling applications. Development of the low-speed turbodrills was determined by the capabilities of the roller-cone bits, or their bearings to be more specific. Another trend in the turbodrilling technology development was the design and construction of DHMs capable of working with cutting-shearing type no-bearing matrix bits, such as natural or artificial diamond, composite inserts, and later PDC bits.

In 1957 and 1958, the first experimental wells were drilled using turbodrills in combination with diamond bits in France and the United States. One of the most important parameters of the efficiency of drilling using diamond bits at significant levels of footage per bit was ROP (v_M) that depended on such factors as rock and bit type, bit rotational speed (n), bit weight (G), and the level of bottomhole cleaning.

The increase of n and G values leads to growth in ROP. For example, the results of experimental drilling, carried out by specialists from the Prikarpatburneft Drilling Company, using electrodrills in combination with 215-mm diamond bits, [24] revealed ROP linear growth at n increasing from 170 rpm to 900 rpm, and G raising from 2 to 17 tons (Q = 23 liters/sec). The energy consumption for the rock destruction process reduced at higher n, whereas it remained constant at increased G. This allowed a conclusion about the expediency of diamond bit drilling application and rotational speed increases of more than 500 rpm.

Yet, the conclusion was accurate only in conditions of perfect bottomhole cleaning. Otherwise, bit weight increase at medium rotational speed levels was more efficient. In turn, quality of the bottomhole cleaning depended on a variety of reasons, such as mud circulation rate Q, pressure differential of the bit, and differential pressure level at the bottomhole.

When using a turbodrill to rotate a bit, the function $v_M = f\ (G)$ cannot be linear and is calculated according to the formula for turbodrill power characteristics (N). It reaches its maximum level at a certain bit weight, the level of which hinges on a specific combination: turbodrill–bit–rock and is determined while drilling.

ROP is dependent on the level of power, applied at the bit:

$$v_M = C_a N^z \qquad 2.28$$

where

z is an exponent, dependent on such factors as bottomhole cleaning quality level, bit weight, and rotational speed

Ca is a constant, dependent on the type of bit

The maximum ROP closely matches the maximum turbodrill effective power N_{ef}

$$N_{ef} = p_t Q \eta_t \qquad 2.29$$

Equation 2.29 indicates that pump pressure and the turbodrill efficiency factor must be increased to achieve high N_{ef} and penetration rate levels.

The requirements of a turbodrill design and the characteristics of diamond bit turbodrilling applications differ slightly from those for regular bit applications. Since diamond bits do not have bearings, their rotational speed is limited only by bit-cutting structure and matrix wear. According to the available information, the limits on rotational speed are quite high, in the range of 800 rpm to 1000 rpm (for 212-mm bits). Requirements for torque in diamond drilling applications are much higher compared to the roller-cone bits. For most types of rock, the specific torque level for cutting-shearing type bits is 2–2.5 times higher compared to the analog values for roller-cone bits. The possibility of drilling at high rotational speeds and the need to increase torque require a significant increase in the effective power of a turbodrill. Construction of a high wear-resistant turbodrill with long service life (that would not limit bit-operating time) could extend this time to hundreds hours.

The existing turbine designs for 195-mm diameter turbodrills did not fully comply with the set requirements, so the 21/16.5 precision cast turbine was developed. Table 2–15 shows the turbine characteristics operating at maximum power and, for comparison, characteristics of other turbines in wells of various depths. For 1.2 g/cm^3 density mud, the pump flow rate was determined to be a constant pressure of 170-180 kg/cm^2 in 141-mm OD drillpipe (DP) with ZSh(FH) type tool joints. A 25-stage rubber-metal seal bearing installed in the Sh2-195 spindle served as a turbodrill axial bearing. Width of the bearing surface was increased by 50% compared to the bearing of the batch produced Sh1-195 spindle, which helped increase the bearing life by reducing specific loads and improving fluid lubrication friction.

TABLE 2–15
Turbodrills of 195 mm OD Characteristics

Turbine Type	Number of sections/ stages in turbodrill	Efficiency factor %	Borehole depth (m) 2000				3000				4000			
			Q	n	M	Δp	Q	n	M	Δp	Q	n	M	Δp
21/16,5 vax casting	2/228	75	33	795	305	105	31	745	270	90	30	725	250	85
21/16,5, vax casting	3/342	75	29	700	355	125	28	675	330	110	26	625	285	95
33/11, vax casting	3/408	56	39	450	355	75	36	415	305	65	34	390	270	60
21/20,5, sand casting	3/288	58	30	660	285	110	29	640	265	105	27	595	230	90
28/18 vax casting	3/330	55	46	390	275	45	41	345	220	35	38	320	190	30
24/18, plastic	3/342	48	43	430	285	60	39	390	235	50	36	360	200	45

Note: M=torque (kg.m), Δp- pressure drop (MPa).

The Saratovneftegaz, Kuibyshevneft, Ukrneft, and Nizhnevolzhskneft Companies carried out commercial test drilling using the 3TSShA-195TL turbodrills. Penetration rates while using this type of turbodrill in combination with diamond bits increased by 30–100% compared to the regular turbodrills. The Archedinsk and Zhirnovsk drilling subsidiaries of the Nizhnevolzhskneft Company tested these turbodrills with diamond bits of Russian and French design made by Christensen Company (bits with differential pressure of 40 kg/cm^2 were used). The drillers used drilling mud with a weight of 1.2-1.3 g/cm^3 (Table 2–15). The bit weight level was selected to achieve the maximum penetration rate and was brought up to 20 tons. For the two-section turbodrills, the optimum bit weight was 8–11 tons, whereas for the three-section turbodrills it was 10–16 tons.

Diamond and PDC bit technologies hold promise for deep turbodrilling applications. Diamond bits meet the requirements of turbodrilling applications perfectly. They operate in a wide range of rotational speeds including high-speed. Introduction of the oil-filled roller-cone bits limited utilization of diamond bits to a certain extent; however, until a special cone seal is designed that enables operation at high rotational speeds, the low speed roller-cone bits should be compared to high-speed diamond bits. The main factor in improving diamond bit economics is the need to achieve higher penetration rate.

An efficient and relatively simple method for increasing diamond bit rotational speed using special turbodrill sectioning was invented and is presented in Figure 2–34. [25]

The upper section shaft (1) was connected to the lower section housing (2). A properly sized end-face collar seal (3), installed in the lower part of the upper section turbine, prevented fluid leak into the annulus under differential pressure in the lower section (2) and in the diamond bit (4). Due to the fact that the stator of the lower section was not fixed but rotated in the same direction as the rotor, the absolute rotational speed of the rotor increased and the stator speed was added to the relative speed of the rotor. The resulting bit rotational speed totaled the sum of the rotational speeds of all shafts in all sections. Unfortunately, the suggested new turbodrill design was only tested once on a test stand. It was never tested in the field and has not been used so far.

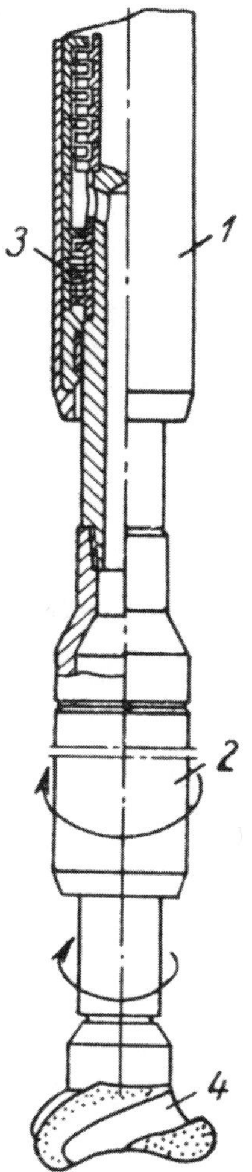

Fig. 2–34 The method for increasing diamond bit rotational speed

Turbodrills for coring applications. Coring using turbodrills with a core barrel attached to the turbodrill shaft did not provide appropriate results because of the high level of vibrations that damaged the core in the core barrel. Therefore, the turbo-coring application was very limited, especially in loose formations of productive horizons. In addition, the limited length of the recovered core sample necessitated frequent tripping operations when coring long intervals using diamond bits.

In 1949, the engineers R. A. Ioannesyan and Gusman designed and built a special turbo-coring unit, called *turbobit*, for wireline coring operations. The KTD3 core turbobit design (Fig. 2–35) was similar to the T12M3 turbodrill. The only difference was a hollow shaft with a wireline core barrel installed inside. The barrel was fitted into the shaft and set in a special tapered surface seat secured inside the turbodrill housing. Hydraulic force was generated by differential pressure in the turbine, and the bit held the core barrel against the seat. The generated friction forces prevented the barrel from rotating. The core barrel had a relief valve that diverted fluid from the inner barrel space to the clearance between the shaft and the barrel when a core entered the barrel. A small clearance of 1 mm between the barrel and the shaft hole and the significant length of the annular slot caused a high level of hydraulic resistance. Therefore, fluid leakage in the annulus was insignificant and did not have a practical effect on turbine operation.

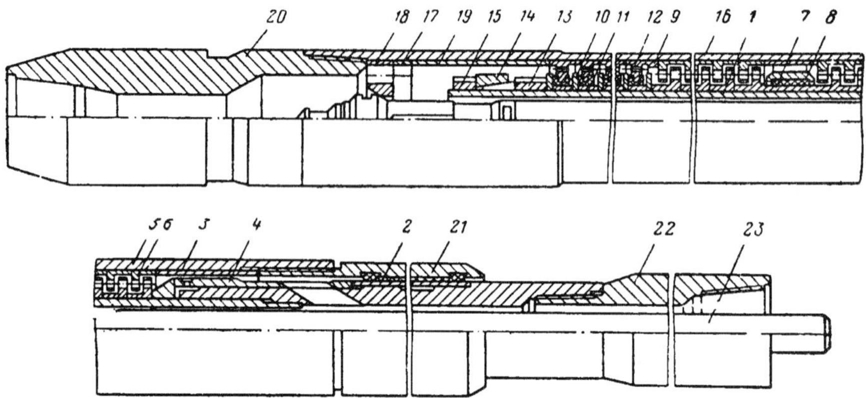

Fig. 2–35 ***The KTD3 coring turbo-bit***

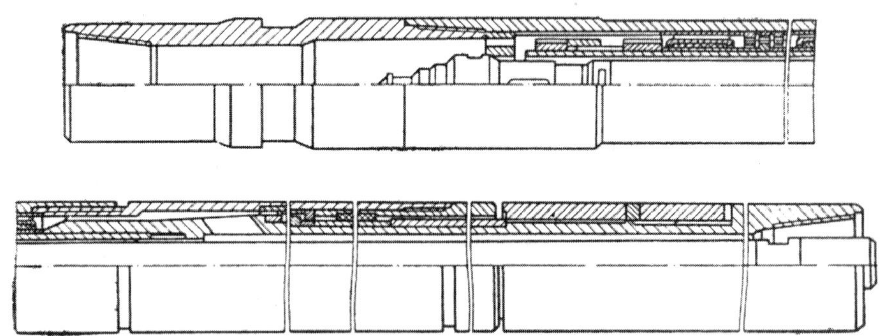

Fig. 2–36 ***The KTD4 coring turbo-bit***

Utilization of the standard turbodrill components for the turbobits, such as a turbine, axial bearings, and radial bearings, enabled recovery of core samples by the KTD3-172 and KTD3-255 turbobits with diameters from 33 mm to 50 mm respectively. The KTD4 core turbobit was later designed and built and was used to recover larger diameter cores (Fig. 2–36). This became possible, thanks to the increase in shaft diameter. KTD4 also had higher torque due to the increased number of turbine stages. Its axial bearing was located at the lower end of the shaft, and its core barrel length could be adjusted depending on the core bit type used.

The industry produced several standard types and sizes of turbobits: KT3-240-265/48, KTD4-195-214/60, KTD4M-172-190/40, and KTD4S-172-190/40. The first digit represented the housing diameter in mm; the two others were for the borehole/core diameter. Unlike the other listed turbobits, the KTD4S-172-190/40 (Fig. 2–37) had two sections, which helped increase torque by increasing the number of the turbine stages. The core length it was possible to recover increased up to 7 m compared to 4 m recovered by a single-section turbobit. The KTD4S-172-190/40 turbobit design was similar to the TS5B two-sectional turbodrill. An axial turbobit bearing, installed in the lower section, took up the hydraulic load from both sections. The housings of the two sections were connected to each other using a tapered thread sub, whereas the shafts were connected through tapered spline couplings. Table 2–16 presents technical and power characteristics of the latest version of core turbobit designs (1980s) for the water circulation regime.

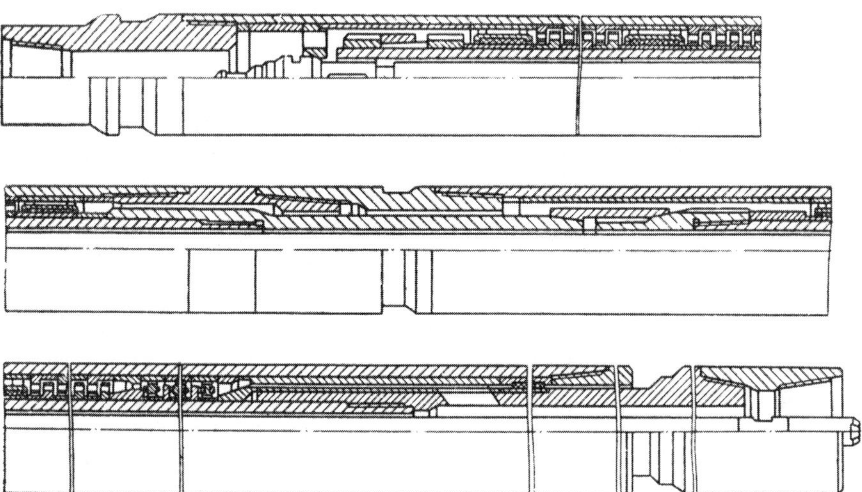

Fig. 2–37 The KTD4S-172-190/40 two section coring turbo-bit

TABLE 2–16
Technical Characteristics of Turbobits for Wireline Coring

TB Type	KTD-195 214/60	KTD4-172 190/40	KTD4-172 2U190/40	2UKTM-172/40
		Turbine parameters		
Length (mm)	10,100	9,080	15,475	18,500
OD (mm)	195	172	172	172
Number of turbine stages	159	135	241	241 floating stators
		Core barrel parameters		
Length (mm)	9,630	8,668	15,067	15,067(1)
Maximal diameter (mm)	95	68	68	65
Core receiver length (mm)	4,635	4,030	7,000	14,000
Core diameter length (mm)	60	40.0	40.0	40.0
Mass (kg)	1,670	1157	2067	2,220
		Turbine power characteristic		
Mud flow (l/sec)	28	28	28	28
RPM	10.4	10.4	8.5	7.5
Torque Nm	1,358.3	1,141.6	1,941.6	1,675
Pressure drop (mPa)	5.7	5.9	9.2	7.2

Turbodrills with floating rotors and stators. One of the promising designs of multistage turbodrills was a turbodrill with floating rotors or stators. The design allowed avoiding a friction-type mounting for one of the turbine stage parts (rotor or stator). In the turbodrills with floating rotors (Fig. 2–38), the rotor transmits rotation to the shaft through a key-type joint or similar type of connection. The rotors sit loosely on the shaft in the axial direction. Each of the rotors with an individual rubber bearing fitted on it is sitting on a corresponding stator. This rubber bearing takes up a limited rotor hydraulic load. Therefore, the level of specific load for each bearing varies from 0.2 to 0.3 kg/cm^2. In conditions of a very low load level, the service life of a rubber-metal bearing of each stage is several hundred hours, even in conditions of highly contaminated drilling mud. A spindle, installed in the lower turbodrill part, takes up hydraulic load, weight load from the sections' shafts, and bit weight.

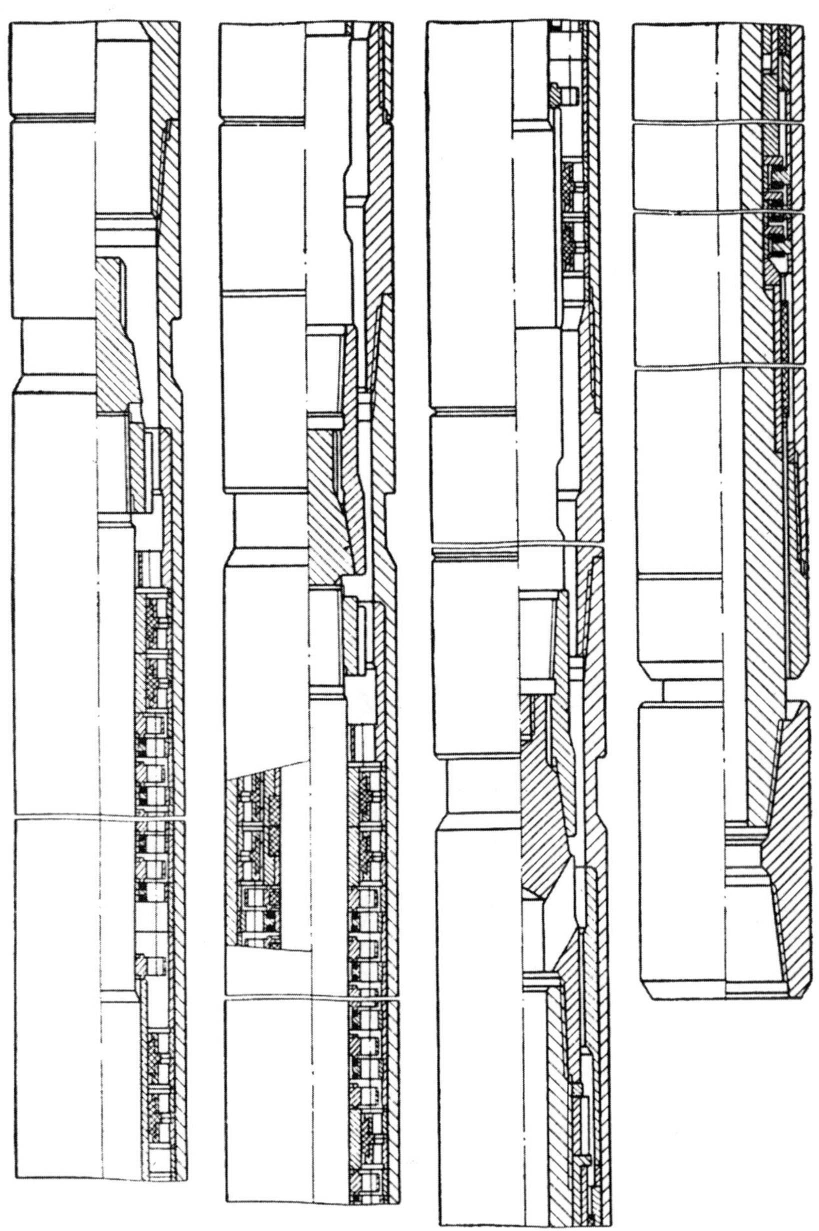

Fig. 2–38 Turbodrill with "floating" rotor

The floating stator features another concept in that type of turbine. The stators fit into the housing by means of the keys on each stator disk and the keyway cut along the internal housing surface. A keyed joint takes up reactive torque from the stator. The turbodrill design with the floating stator led to a considerably increased average diameter in the turbine, thanks to the absence of the stator thrust faces, which significantly improved the turbine characteristics.

Bearings installed in each turbine stage sufficiently reduced axial clearances between the rotor and the stator. In regular turbines, axial clearances are normally 14–20 mm compared to 5–8 mm in the turbodrills with floating working elements. As a result, the number of turbine stages in a section could be increased significantly. One of the main advantages of the floating working elements design was that they do not require an adjustment of the turbine axial clearances, thus making it possible to increase the number of turbodrill stages.

Figures 2–39, 2–40, and 2–41 show a general view of a turbodrill with a floating stator, a turbine stage, and a stator position inside the turbodrill housing. Table 2–17 presents characteristics of turbodrills with floating stators and rotors.

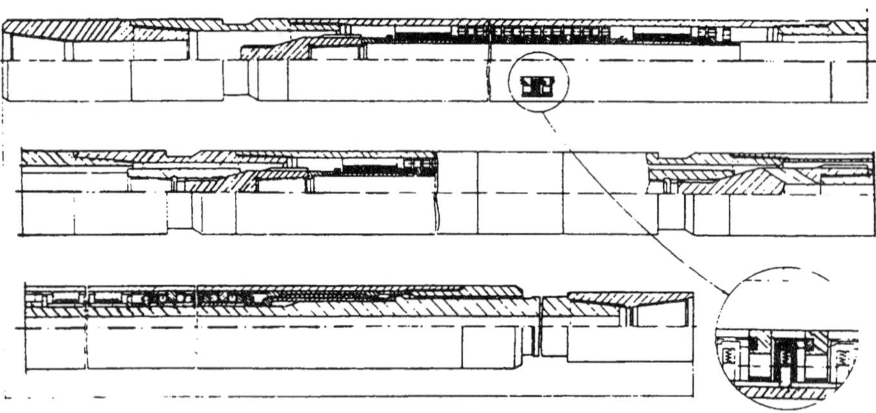

Fig. 2–39 Turbodrill with "floating" stator

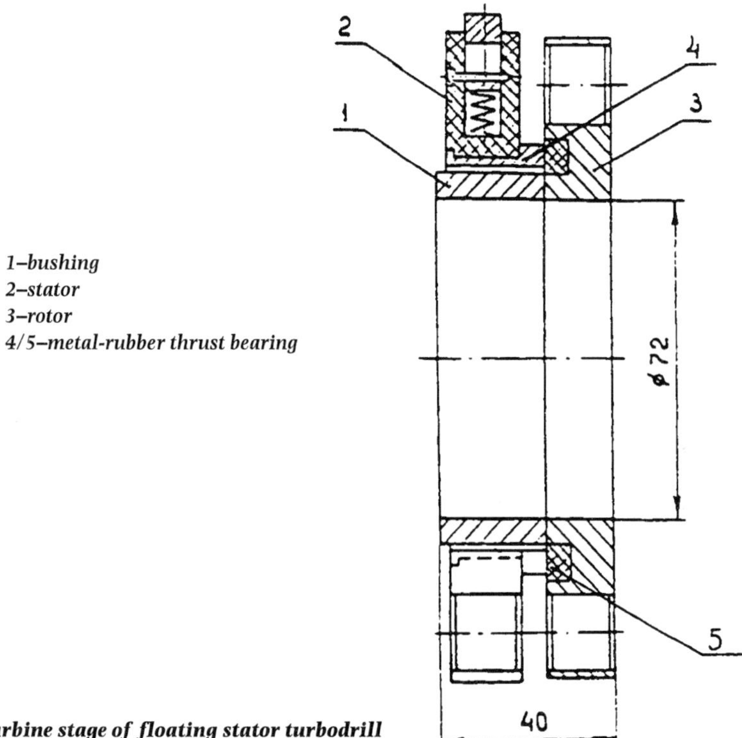

1–bushing
2–stator
3–rotor
4/5–metal-rubber thrust bearing

Fig. 2–40 Turbine stage of floating stator turbodrill

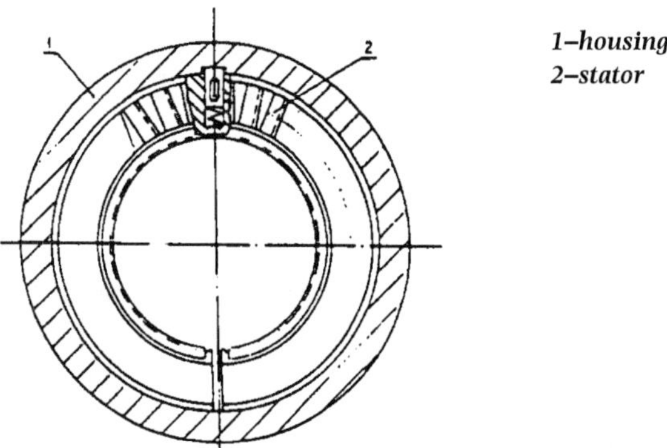

1–housing
2–stator

Fig. 2–41 Floating stator position in turbodrill section housing

TABLE 2–17
Turbodrills with Floating-Type Stators and Floating-Type Rotors (3T-105K) Characteristics (Drill Mud–Water)

						Parameters at operating mode			
Turbodrills type	OD (mm)	No. of turbine stages	No. of turbine sections	Length (mm)	Weight (kg)	Flow rate (l/sec)	RPS	Torque (kg m)	Pressure drop (MPa)
TSSh-1M1-195	195	425	3	25,800	4,200	28	5.4	1,788	3.8
TPS-195M	195	380	2	18,400	3,085	32	6.2	1,563	3.3
TPS-172	172	426	3	26,250	3,325	25	6.7	1,697	4.8
TPS-172M	172	432	2	18,500	2,180	25	83	1,668	6.8
TPS-105	105	420	3	15,500	765	12	10	412	5.4
3T-105K	105	303	3	12,700	590	12	11.6	430	6.8

A comparison between the data shown in Table 2–17 and the information from Table 2–9 (showing characteristics of three-section turbodrills including those with precision cast turbines) indicates that the development of turbodrills with floating stators is another step toward the improvement of *M*/*n* (torque/rpm) ratio, i.e., the dynamic characteristics of turbodrills. This conclusion can be made after comparing the turbodrill designs presented in Table 2–18.

TABLE 2–18
Sectional Turbodrills Characteristics as Compared to Floating Stators Turbodrills (Operational Mode)

Turbodrills type	Turbine type	No. of turbine sections	No. of turbine stages	Flow (l/sec)	ROP	Torque (kg m)	Pressure drop kg/cm^3
TSSh-1M1-195	Floating stator	3	425	28	324	178.8	38
TPS-195M	Floating stator	2	380	32	372	156.3	33
3TSSh-195	Serial	3	306	30	400	130	35
TPS-172	Floating stator	3	426	25	400	169.7	48
TPS-172M	Floating stator	2	432	25	498	166.8	68
3TSSh1-172	Serial	3	336	22	555	120	73

The TSSh-1-M1-195 turbodrills with floating stators were field-tested in Glavtyumenneftegaz, Western Siberia. The test results indicated a significant improvement of bit performance (2 to 3 times) and increased MTBF compared to the standard turbodrills. This type of turbodrill was also tested by the Kuibyshevneft Company, while drilling experimental well No. 163 (this drilling test is described in Chapter 3).

In 1982 and 1983, the Kaliningradmorneftegaz Company carried out a large amount of drilling work using this type of turbodrill to drill 18 wells in various fields offshore. [26] An interval was drilled below the intermediate casing into the Cambrian deposits from depths of 1100–1200 m to the TD at 2300 m. The drilling mud density was 1.18—1.2 g/cm^3. Jet bits with a differential pressure of 3 to 5 MPa drilled a total of 7000 m in 1800 hours. The results of this test drilling with TSSh-1-M1-195 floating stators turbodrills revealed for tri-cone bits application an increase in the penetration per bit by 59%, and ROP by 24% in comparison with standard drilling that used the A7Sh turbodrills (mostly utilized by the company).

During the test drilling, seven ISM-212-type bits (ISM means drag type bit with diamond composite inserts manufactured by Institute of Super-hard Materials in Kiev, Ukraine) were also used to drill 2140 m during 840 hours. The average footage per bit was 305 m at an average penetration rate of 2.55 m/hr. The average bit on bottom time was 120 hours. No cases of the turbodrill failure were registered. The average service life of turbines in the tested TSSh-1-M1-195 turbodrills was 550 hours. The MTBF of the ShShO-type spindle section with a roller bearing with rubber absorbers was 277 hours compared to 85 hours for a regular turbodrill. These improved parameters were especially important for offshore drilling applications.

Based on the test results, the following are some conclusions about the advantages of turbodrills with floating stators compared to regular sectional turbodrills:

- Improved power characteristics due to the larger average diameter of the turbine and an increased number of stages in each section.
- Elimination of the most complicated and labor consuming operations from the turbodrill assembly and adjustment process.
- Longer spindle MTBF, thanks to the elimination of the limits on allowable axial backlash of a spindle bearing.
- Longer turbine section MTBF brought about by the elimination of the wear on the blade cascade by fitting the rotor on the stator.

Shifting to the wide utilization of the new design turbodrills was, undoubtedly, one of the most challenging and rewarding tasks for engineers and drilling companies. Unfortunately, this task has never been fulfilled. The design features of turbodrills with a floating stator/rotor required a significantly different manufacturing procedure and different materials. Because the use of rubber and plastic was a mandatory requirement for achieving advanced characteristics in the turbine stage design, operational temperatures have limited the turbodrills with floating stators application.

Geared turbodrills—new winds of development

Design attempts in 1960s and 1970s—promising tests. An analysis of the methods of improvement of the sectional turbodrills characteristics just reviewed made it evident that the lowest allowable rotational speed in the turbodrill shaft is 250–300 rpm, which was proven in operations. However, both rpm and torque were not sufficient to drill deep wells efficiently using roller-cone bits. Therefore, scientific, research, and design organizations have addressed the issue of the development of gear-reduction turbodrills once again using new achievements in mechanical engineering.

In the 1960s and early 1970s, several research institutes, such as the VNIIBT and Oil and Gas University (MINKh or MING) in Moscow, the Perm Branch of the VNIIBT, and the Kungur Engineering Plant, developed a number of gear-reduction turbodrills. The prototype tests revealed that engineers had not designed a sufficiently reliable oil-filled reduction gear or durable non-sealed gear.

However, some positive results were obtained from the experience gained in the construction of reduction gear inserts for electrodrills (discussed later in this chapter). The SKTBE of the Ministry of Electrical Engineering Industry of the FSU developed reduction gear inserts for electrodrills of all standard sizes with gear ratios of 2, 3.14, and 9. When used in Turkmeniya and Bashkiriya, they helped increase footage per drillbit and penetration rate by 15–20% compared to non-reduction gear electrodrills. [27]

As a part of the electrodrill oil-filled system, the gear insert located between the motor and the spindle operated in most favorable conditions. This ensured the high operating efficiency of the electrodrill. Using these developments, the VNIIBT in conjunction with SKB, developed and built a spindle-type reduction gear turbodrill with the RT-195 attached reducer (Fig. 2–42). The reducer had a double-row planetary gear (gear ratio 2.92) with straight involute gearing. A bevel gear (differential), working as a torque divider, distributed torque between the rows.

The oil-filled reduction gear system included a no-inertia diaphragm lubricator. A stack of radial rubber collars and sequentially installed end-face seal bearings with hard-alloy rings were used as seal elements. The reducer shaft was connected to the ShSh 01-195 spindle (discussed in this section) through a spline coupling with an independent lubrication system. The reduction gear with the spindle was connected to one, two, or three turbine sections. Table 2–19 presents power characteristics of this type turbodrill for mud with a weight of 1.8 g/cm^3. Using tachometers, the Kaspmorneft Company tested this turbodrill in wells with depths of more than 3000 m and had stable turbodrill operations at 130–170 rpm. ROP was higher compared to rotary drilling.

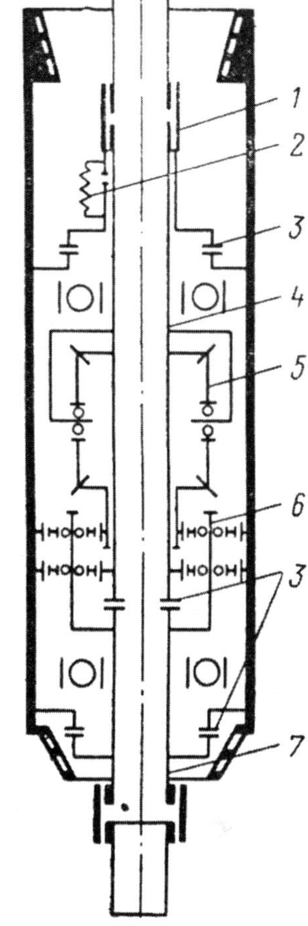

Fig. 2–42 The RT-195 spindle-reducer of new geared turbodrill (1960s)

1–coupling
2–lubricator
3–seals
4–shaft
5–gear differential
6–carrier
7–low speed shaft

TABLE 2–19
Turbodrill with Planetary Gear Insert RT-195

Number of Turbine Sections	Turbine type	Mud flow rate (l/sec)	RPM	Torque (kg m)	Pressure drop kg/cm^2
1	33/11	34	130	410	30
2	33/11	32	110	720	52

The turbodrill reducer insert with the stack type oil-filled system was the most promising development of the turbodrill for deep well drilling applications. It used regular seals and special labyrinth type seals.

In 1970, specialists from the Kungur Engineering Plant and engineers from the VNIIBT Branch in Perm developed the TR2Sh-195 non-sealed reduction gear turbodrill. The turbodrill consisted of two turbine sections and one axial bearing section (spindle). A multi-row planetary friction type gear reducer was installed in each turbine section. Each row of the reducer consisted of the turbine stator and rotor discs with races and balls. The balls were located in a separator connected through a key-joint to a section shaft. The load, generated by a differential pressure in the rotor, provided the pressing force required for torque transmission. Since the reducing gear used the normal turbine circulation degree of a regular turbodrill, the pressing force of the friction couple was not dependent on the transmitted torque level and was practically constant in any operating regime.

Table 2–20 shows characteristics of the TR2Sh-195 turbodrill with the regular turbine 21/20.5. [28] In conditions with no sliding, the reducer gear ratio was 2–7. However, in conditions of vibration, there was always a risk of sliding, which resulted in a decrease in the dynamic gear ratio at low rotational speed. The TR2Sh prototype field tests indicated that the service life of the main gear elements, such as balls and their bearings in carriers, was about 50 hours. [29]

TABLE 2–20
Characteristics of the TR2Sh-195 Turbodrill with Ball-Type Gear Reducer (Turbine Type 21/20.5)

Number of Turbine Sections	Number of Turbine stages	Mud flow rate (l/sec)	RPM	Torque (kg m)	Pressure drop kg/cm²
1	58	45	290	250	49
2	116	40	258	258	77
		35	225	225	60

In the other designs, the RSh-195 non-sealed friction type gear reducer, the same type gear was built as a separate spindle and located between the turbine and the bearing (Fig. 2–43). The reducer was built according to the multi-row scheme with a joint carrier. To provide equal distribution of the friction gear pressing force against the rows, the reduction gear assembly was rolled on under load in abrasive media or special compensating elastic elements were used. The gear reducer kinematic gear ratio was 2.5. Depending on bit operating conditions, the gearbox mechanical efficiency factor varied from 0.55 to 0.7, which resulted in significant variation of the dynamic gear ratio. The gear reducer tests indicated a low operating life of 40–50 hours in abrasive mud because of inadequate durability of the solids of revolution (steel rolling bodies) and their bearings in the carrier.

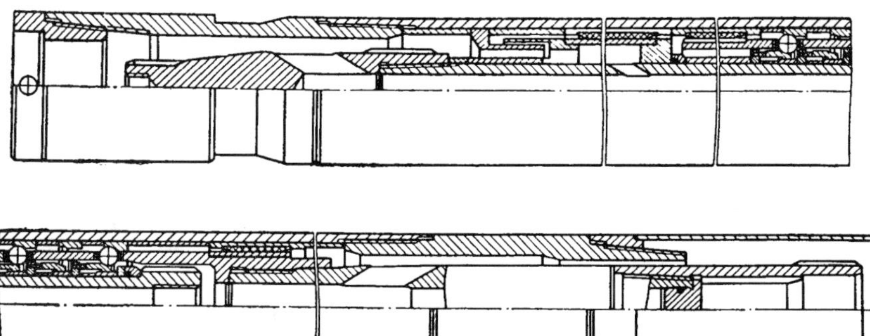

Fig. 2–43 Reducer-spindle RSh-195 of planetary friction-ball type reduction gear turbodrill (1970)

However, continued research work in this field has confirmed good prospects for developing oil filled gear reducers with enhanced seal design. Engineers from the Perm Branch of VNIIBT (N. D. Derkach, E. N. Krutik, et al.) designed and built the RM-195 gear reducer, and its tests produced positive results. [30] This appeared to be the first commercial reduction geared turbodrill.

TRM-195 gear reduction turbodrill application—first commercial success. From 1975 to 1978, the Perm Branch of VNIIBT pioneered the development of the 195-mm diameter gear reduction turbodrill, TRM-195, with oil-filled gear reducer RM-195. [31] Commencing in 1978 through 1985, this turbodrill was field-tested in Bashkiriya and Tatariya, Western Siberia and in the ultra-deep wells drilled in the FSU according to the Scientific Continental Drilling Program (*see* Volume 2).

A diagram of this turbodrill is shown in Figure 2–44. The TRM-195 includes the turbine section (A), top and bottom spindles (B, D) with the oil-filled gearbox (C) mounted between them. The gearbox contains input and output shafts installed on bearing assemblies and the planetary gear positioned between the shafts. The gear protection system consists of the two face seals and a lubricator to transfer the mud pressure inside the oil-filled chamber that incorporated the gear and bearings. The shafts are interconnected with the spindle shafts by virtue of clutches. The gear reducer transforms the rotation of the turbine section shaft and outputs it to the spindle shaft and drillbit while reducing the rotating speed proportionally to its gear ratio (3–69) and increasing the torque proportionally.

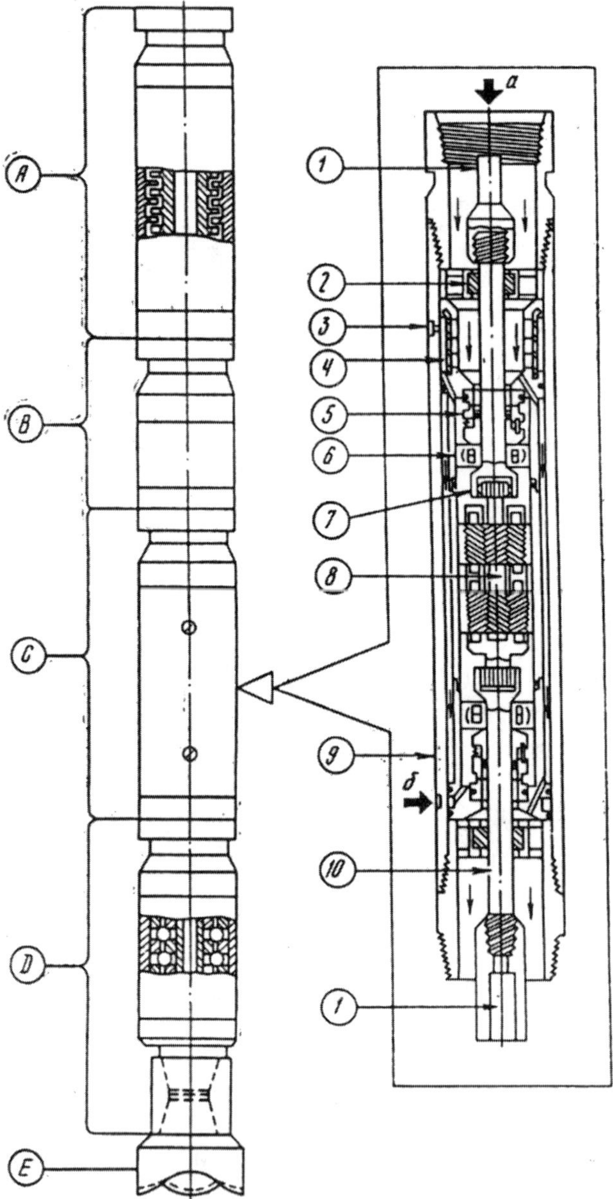

A– turbine section
B–thrust bearing spindle–intermediate
C–reduction gear box
D–bottom spindle
E–drill bit

The box (c) contains input 7 and output 9 shafts installed on rolling bearings 6 and radial sliding bearings 2. Planetary gear 8 positioned between shafts 7 and 9; oil protection system consisting of two face seals 5 and lubricator 4 to transmit mud pressure inside oil-filled chamber incorporating gear and bearings. Mud passes through circular port inside housing 10 (direction is shown by arrow). Oil is filled through orifice 3 in housing 10 being plugged. The shafts 7 and 9 are interconnected with spindle shafts by virtue of half-clutch 1.

Fig. 2–44 The TRM-195 turbodrill with planetary gear reducer

The production of oil-filled gear reducers RM-195 in quantity started in 1986 at the Polymermash plant in Tambov with an annual output of 100–230 sets. The main customers for the TRM-195 were the drilling enterprises of Western Siberia. Between 1987 and 1991, more than 10 service centers for maintenance of the gear reduction turbodrills were organized in oil companies like Noyabrskneftegas and Nizhnevartovskneftegas. In 1989 and 1990, the volume of drilling using the gear reduction turbodrills increased from several tens of thousands of m to 500,000–600,000 m annually. TRM was used most effectively in the hard formations below 2000 m depth. Since 1987, more than 3.5 million m have been drilled.

The heat resistant (250–300° C) TRM-195 version enabled the drilling of the Kola ultra-deep well to a record depth of 12,262 m. The TRM-195 turbodrill was successfully used in drilling the Saatlinskaya (Azerbaijan), Uralskaya, Tyumenskaya (Russia), Krivorozhskaya, and Dneprovo-Donetskaya (the Ukraine) ultra-deep wells.

Different types of turbines may be used in the gear reduction turbodrill TRM-195 (*see* data in Table 2–21). The best ROP was achieved when using assemblies two and five as specified in this table. In the intervals of 2000–3000 m at the Western Siberia fields, the TRM turbodrills allowed drillers to maximize footage per bit (Russian production) by 30–50% and trip speed by 20–30% compared to the regular turbodrills. Average MTBF of the gearbox in conjunction with the two spindles constituted 120–130 hours (per data from Noyabrsk Service Center). The most typical reason for failure in this case was spindle wear. Average overhaul life of the RM-195 oil filled gear was 170–220 hours; however, sometimes the failure interval exceeded 300 hours. The average guaranteed failure interval of RM-195 until write off constituted 900 hours but this procedure usually took place after 2000 hours of operation.

TABLE 2–21
TRM-195 Gear Reduction Turbodrill Characteristics with Different Turbine Types

						Operating Conditions			
Assembly Number	Turbine type	No. of turbine sections	Gear ratio	Mud flow rate (l/sec)	M_{op} (N.m)	n_{op} (Mpa)	P_{op} (min^{-1})	M_{st} (N.m)	N_{n-1} (min^{-1})
1	TSSh1-195	2	3.69	32	3,720	111	4.4	7,440	222
2	TSSh1-195	3	3.69	32	5,580	111	5.9	11,160	222
3	A7Sh	1	3.69	32	3,300	161	6.0	6,600	322
4	TSShA-195TL	1	3.69	32	3,780	206	5.4	7,560	412
5	TSShA-195TL	2	3.69	24	4,252	155	6.1	8,504	310
6	TSShA-195TL	0.5	3.69	2	1,890	206	3.8	3,780	412

TRM-195 with drillbits from USA—operational experience. The oil production companies in Russia had a chance to use high quality foreign-made drillbits in the 1990s. Despite the high price of the rolling cutter bits from such companies as Reed Tool, Smith International, and Hughes Christensen Co., they proved their efficiency in the lower borehole intervals of the Western Siberian fields. High quality bit bearings could run at the bottom at up to 200 rpm for 70–100 hours in average. Table 2–22 represents field examples of drillbit performances with different motors, and Table 2–23 represents the average results of PDM and TRM motor applications with the advanced drillbits used in Western Siberia fields during 1994 and 1995.

TABLE 2–22
Drillbit Records, West Siberia, 2000–3000 m Deep Sections

Oil Company Contractor	Motor type	Bit specs.	Bit manufacturer	Number of bits	Avg. penetr. (m)	Bit life, (hr)	ROP, (m/hr)
UKOS, NUBR-1	D2-195	S83F	Dresser Security	1	565	135	4.2
		EHP-51A	Reed Tool Co.	1	522	76	6.9
		MF-15	Smith Int.	2	782.5	115	6.8
		215.9 MS-GNU	Volgoburmash	170	83.2	16.8	5.0
UKOS, NUBR-2		MF-15	Smith Int.	11	612.6	15.6	5.8
		215.9 MS-GNU	Volgoburmash	50	83.7	16.4	5.1
UKOS, MUBR		EHP-51A	Reed Tool Co.	1	774	96	8.0
		ATM-P-11H	Hughes	1	301	47	6.4
		MF-15	Smith Int.	10	651.8	140.4	4.6
		215.9 MS-GNU	Volgoburmash	168	83.3	19.8	4.2
UKOS, SUBR		ATM-P-11H	Hughes	1	917	73	12.6
		MF-15	Smith Int.	12	653.9	75.4	8.7
		215.9 SGV-2	Volgoburmash	98	97.7	6.7	14.6
SLAVNEFT, Megionneftegas		MF-15	Smith Int.	2	559.5	93	6.0
	D1-195	MF-15	Smith Int.	1	365	80	4.56
Povkhovskoye UBR	D2-195 (84%) TRM-195 (16%)	MF-15	Smith Int.	11	596.1	99	6.02
Mirnensky UBR	TRM-195 (76%) 2xD2-195 (24%)	MF-15	Smith Int.	13	742	94.4	7.86
Purneftegas	TRM-195	MF-15	Smith Int.	6	658	103.9	6.33
		S83F	Dresser Security	3	674	85.5	7.88
		ATM-P-11H	Hughes	1	436	38.5	11.3
SIBNEFT, Noyabrskneftegas SUBR-1	TRM-195 (81%) D2-195 (19%)	MF-15	Smith Int.	14	478.5	44.2	10.8
Kholmogorskoye UBR	TRM-195	ATM-P-11H	Hughes	17	664.2	79.8	8.32

TABLE 2–23
U.S. Drillbit Performance with Russian Downhole Motors, West Siberia 1994–1995

Bit specs.	Bit manufacturer	Motor type	Number of bits	Average penetration (m)	Average bit life (hr)	ROP (m/hr)
MF-15	Smith International	D2-195	48	632.2	103.6	6.10
MF-15	Smith International	TRM-195	33	614.9	74.8	8.22
ATM-P-11H	Hughes Christensen	TRM-195	18	651.5	77.5	8.41

Efficiency calculations were made of ATM-11H (Hughes Christensen $8\frac{1}{2}$-in. diameter bits). A comparison was made to the Russian SGV, R45, R54, and MZGV bits used in the interval between 2000 and 3100 m. The price of the Russian bit was in the range of $700 to $2000 USD, depending on the class. Average drilling rig costs were approximately $300 USD per hour.

Average ATM-11H bit performance was as follows:

- the average footage per bit was 632.5 m and exceeded the average footage of Russian bits by a factor of 9 times;
- the average bit life was 76.1 hours and exceeded the life of Russian bits by a factor of 7.7 times;
- the cost to drill one meter was $49.90 USD, a reduction of 27.6% compared to Russian bits.

Drilling the 2000–3100 m interval with the Hughes Christensen bit saved about $12,064 USD. Based on experience, it was determined that the best drive for these bits in the conditions of Western Siberia was the gear reduction turbodrill. The life and reliability of the TRM-195 turbodrill could run the bit completely to the bottom, and the stability of the turbodrill provided optimal drilling throughout the entire run. [32]

Development of new generation gear reduction turbodrills. Throughout the time it took to master production and increase drilling volumes using the gear reduction turbodrills, massive work was done to improve their design—taking into account the drilling conditions. The turbodrill on the bottom featured the generation of substantial dynamic loads due to longitudinal, torsion, and lateral vibrations of the lower section of the DS. These conditions brought up the problems of dynamics and hydrodynamics and the need for comprehensive study and evaluation of the dynamic processes to assure the reliability of the gear reduction turbodrill.

For several years, the Perm Branch of VNIIBT and later JSC Neftegaztechnika, in cooperation with a number of other research companies in Perm, performed theoretical and experimental studies of DHM dynamics. Mathematical models of a downhole drive with a gear reduction and oil protection system were developed. Model calculations improved the primary part of the double-row planetary gear using Novikov's helical gearing. Protection for the gear and satellite bearings from overloads and impact of WOB vibrations was introduced into the design. The oil protection system was improved significantly. The following were included in these studies:

- calculation of the face seal critical mass including hydraulic resistance
- calculation of the effect of hydrodynamic vibrations on the reliability of the face seal
- determination of an optimum ratio of hydro-discharge coefficients of the top and bottom seals

This resulted in the development of design and production technology of the seals to assure their reliability in an abrasive environment at intensive axial and radial vibrations and mud pressure pulsation. Methods to increase the reliability of gear reduction and extend the failure interval to 500–600 hours were developed as well. Technological equipping of production shops and improved quality of manufacturing were required in addition to the high level of design needed to handle the dynamic loads. [33]

This work formed the basis of the development of a new generation of gear reduction turbodrills. Since the overhaul period of the RM-195 oil-filled gearbox was twice as large as the overhaul period of the top and bottom spindles that worked in mud, it was natural to mount axial bearings in the oil-filled chamber of the gearbox. This new assembly, called the gear reduction spindle RSh, was connected to a serial production turbodrill instead of a conventional spindle at the site. It could be used in one, two, or even three turbine sections with the gear reduction spindle—depending on drilling conditions and the power required for the bits used. A diagram of the gear reduction turbodrill with RSh spindle is shown in Figure 2–45.

The 195-mm gear reducer spindle of the design previously mentioned was developed in 1987. It was field-tested in the Perm Region and Udmurtiya, and quantity production started in 1992. The overhaul time for the gear reduction

Oil-filled chamber of the gear reduction-spindle protected by top seal 4 and bottom seal 5 contains axial bearings–top 6 and bottom 7, planetary gear reduction 8 and radial bearings 9. High-speed shaft 10 is interconnected with turbine shaft 11 by virtue of bevel spline clutch; bit 3 is connected to low-speed shaft 12

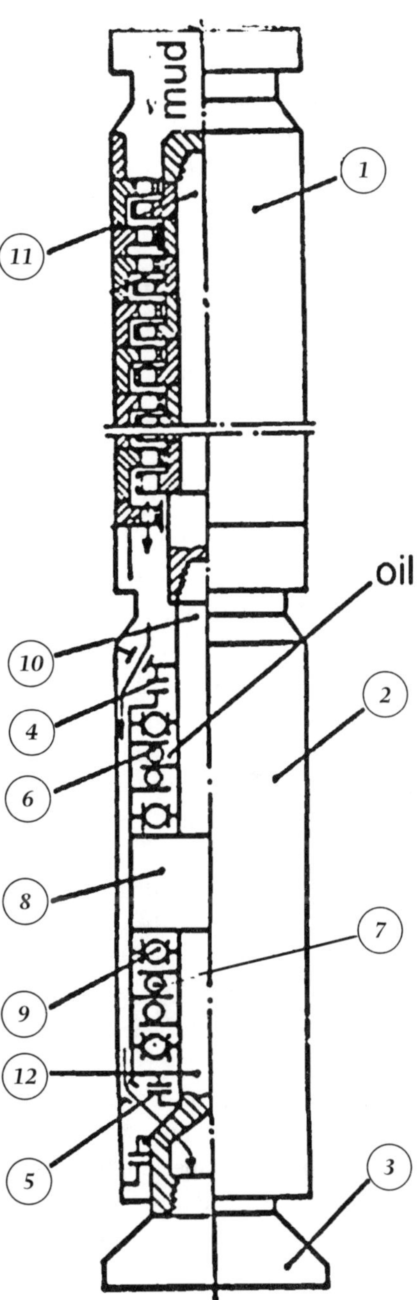

Fig. 2–45 The TRSh type gear reduction turbodrill

spindle was about 200 hours. The RSh-105, RSh-145 (or -142) RSh2-195, RSh2-240 gear reduction spindles were developed and manufactured analogous to the design previously described. At present, the RSh2-240 gear reduction spindle has two more modifications: RSh3-240 and RSh5-240. Gear reduction spindle characteristics are shown in Table 2–24. The TR-145 was designed based on TRV-142 experience.

TABLE 2–24
Turbodrills with RSh-Type Gear Reduction Spindle Characteristics

Turbodrill assembly	OD (mm)	Length (M)	Flow rate, (l/sec)	torque (N.m)	Operating RPM	Pressure drop (MPa)	Poer (kW)
RSh-105 + two TSSh-105 sections	105	2,7	8	1,040	250	7,3	14
RSh-145 + TRV-142 section	145	2,9	22	3,500	220	5,0	40
RSh2-195 + two TSSh1-195 sect.	195	4,8	34	10,100	210	6,0	65
RSH5-240 + TSSh1-240 section	240	3,9	45	11,700	205	6,1	120

The TRV-142 reduction gear turbodrill was designed to work in combination with retractable bits without pulling out the DS (*see* Table 7–4 in Chapter 7 of Volume 2) and was developed in accordance with the RSh principle. This type of turbodrill was successfully used for drilling the Krivoy Rog super-deep borehole in the Ukraine. The important feature of the reducing spindle of this design was that the flow bypassed the gearbox outside the housing, which provided a very reliable seal design. In Volume 2, Chapters 5 and 7 contain more information about this application.

Advanced geared turbodrills for horizontal drilling. Since directional and horizontal drilling found wide application around the world in the 1980s and 1990s, new requirements were imposed for DHMs. Short length motors with a bent in the lower section were needed. Both high rpm and torque were required to drill horizontal well intervals with diamond and PDC bits.

Another generation of gear reduction turbodrills was developed to do this. The design concept of the new gear reduction turbodrills was based on the combination of a super high-speed turbine (1800–4000 rpm depending on turbine diameter) and a reducing planetary gear with a gear ratio of 3.4 to 3.6, which provided the needed power. The turbine was selected from the newly developed turbodrills that followed improved drillbit designs and technology demands (*see* the next section in this chapter). Four standard sizes of TR-type turbodrills had been developed—120-, 178-, 195-, and 240-mm OD. The market demanded the development of an 88- to 95-mm OD gear reduced turbodrill as well. The 120-mm gear reduction turbodrill was designed jointly with the German company TIEBO Tiefbohrservices within the framework of the "Thermie" European Commission project under the Contract OG-201-98: "Hot Resistant Gear Reduction Turbodrills for Efficient Exploration of Hydrocarbons" in 1998 and 1999 and presented at Offshore Technology Conference in 2000. [34]

The designs of 120-mm and 178-mm (TR2-120G and TR2-178G) featured turbine-gear reduction and spindle sections interconnected by a bent sub. The drive shaft and planetary gear were mounted in the upper section. The following parts were assembled on the shaft: multiple stages of high-speed turbine (up to 100), an oil protection system, and a multi-row toroidal bearing assembly. The drive shaft radial bearings were faced with hard alloy materials and were reliable at increased rotating speeds and in various drilling mud environments.

The spindle section included the radial bearings, multi-row axial bearing, and lower face seal oil-mud. To reduce mud leakage along the spindle shaft, a longer hard alloy radial bearing was used, and the clearance created slot ring packing. The transfer of rotation from the planetary gear drive to the spindle shaft was provided through an intermediate shaft designed to allow no more than two degrees of axes misalignment. All the units worked in the common environment of the oil-filled chamber, including the planetary gear, drive and driven shafts, support bearings, and intermediate shaft.

The 195-mm and 240-mm gear reduction turbodrills were the simplified version of this design. The turbine section had 140–150 stages of high power turbines with independent suspension on the separate axial bearing assembly. The turbine section was joined with the gear reduction spindle section by virtue of cross clutches working in the open mud environment. Toroidal axial bearings were replaced with special roller thrust bearings of increased load capacity. In other respects, the design was analogous to that previously described.

The TR2 turbodrills were given the most efficient power characteristics and operated in the range of 260 to 470 rpm depending on the diameter and flow rate. These parameters relate favorably to modern PDC drillbit applications in different well environments, including hard rock and high temperatures.

In a number of cases, sealed cone bits were used to drill directional and horizontal wells, which required moderate rotating speeds of less then 200 rpm. The TR3-120G, TR3-178G and TR3-195G gear reduction turbodrills were equipped with double stage planetary gears to meet this requirement. Half the turbine stages were installed in the turbine section, and the total turbodrill length was reduced to drill high build rate intervals with sufficient turbodrill power. Parameters of the TR-type turbodrill are given in Table 2–25. These turbodrills were produced conventionally to be used at bottomhole temperatures up to 150° C and to be heat resistant in applications up to 300° C.

TABLE 2–25
Characteristics of Gear Reduction Turbodrills for Directional and Horizontal Wells Drilling (Mud Density—1,2 g/cm³)

Turbodrill code	Diameter (mm)	Length of spindle (m)	Length of turbine (m)	Flow rate (l/sec)	Pressure (MPa)	RPM at free rotation (min^{-1})	Stall torque (N.m)	Output power (kW)
TR2-120G	120	1.65	5.35	12	54	820	2,010	39
				14	74	956	2,730	62
TR3-120G	120	1.65	4.55	10	16	217	1,840	10
				12	24	261	2,640	16
TR2-178G	118	1.56	7.2	23	77	574	5,860	88
				27	106	674	8,070	142
TR3-178G	178	1.56	4.4	27	35	175	10,380	47
				30	44	194	12,820	65
TR2-195	195	2.56	7.2	28	87	480	7,100	80
				30	100	523	8,150	99
TR3-195	195	2.56	5.2	30	38	142	11,280	37
				32	43	151	12,840	45
TR2-240	240	2.10	5.2	70	53	531	16,860	234

The new generation of gear reduction turbodrills was field-tested in Russia, Germany, the United States, Canada, Venezuela, Egypt, and Saudi Arabia. The test results demonstrated that it was possible to develop turbodrills to meet the requirements of modern bit drives and achieve improved drilling results.

Turbodrilling experience in the 1990s—a new challenge

Modern trends of turbodrill design, development, and application. During more than 50 years in the development and use of commercial turbodrilling methods, research and design engineers have been striving to resolve the main issue related to the level of worldwide competitiveness. They applied themselves to designing a bit drive for optimum borehole deepening without the limitations of durability.

The existing drilling techniques had at least two mutually exclusive DHM design and power characteristics used with drillbits of essentially different designs. Directional and horizontal drilling applications increasingly dictated the necessity of developing a drive with the smallest possible axial dimensions for both low rpm roller-cones and high rpm PDCs and diamond drillbits. Theoretically and practically, such a design solution had long been developed, that is, a relatively short turbodrill with an oil-filled gearbox or a spindle type reducer (described in the previous section). The fact that the wide-scale production of a reliable gear reduction turbodrill remained problematic prompted design engineers to develop alternate solutions.

A number of designs for low speed turbodrills were available in Russia. Those designs controlled the rpm of the turbodrill shaft by using such techniques as combinations of different speed turbines, a turbine and a PDM drive (screw pair) on one shaft—compound motor, and installing a mechanical multiplication device. Still, none of the existing designs fully resolved the problem since the rpm was controlled at the expense of a reduction in power. Moreover, they did not significantly reduce the motor length. When a compound motor (Fig. 2–46) was used for rpm control, the turbodrill drawbacks were typical for PDMs, such as limited operating temperature (120° C) and durability.

Without a comprehensive solution, these systems were considered feasible given the level of industry development at the time. Table 2–26 presents power characteristics of the compound motors used during the last few years using standard turbine sections at a maximum diametric clearance of 0.3 mm in the PDM section. [35]

TABLE 2–26
Turbine-PDM Compound Motors Characteristics

	Downhole motor external diameter, mm		
	172	195	240
RPM	60–110 160–250	70–120 160–260	110–170 240–300
Torque (N.m)	2,400–5,000	3,000–6,500	6,000–11,000
Flow rate (l/sec)	18–24	20–28	30–40
Pressure drop (MPa), mud SG 1100 kg/m^3	6.9–9.1	6.5–8.7	6.2–8.5
Length (m)	13.5 (20.8)	13.5 (20.8)	13.5 (20.8)
Mass (kg)	2,010 (3100)	2,600 (3950)	3,900 (6050)
Bit diameter recommended (mm)	190.5–215.9	215.9–244.5	269.9–363.7

The length and mass for compound with two turbine sections are indicated in brackets.

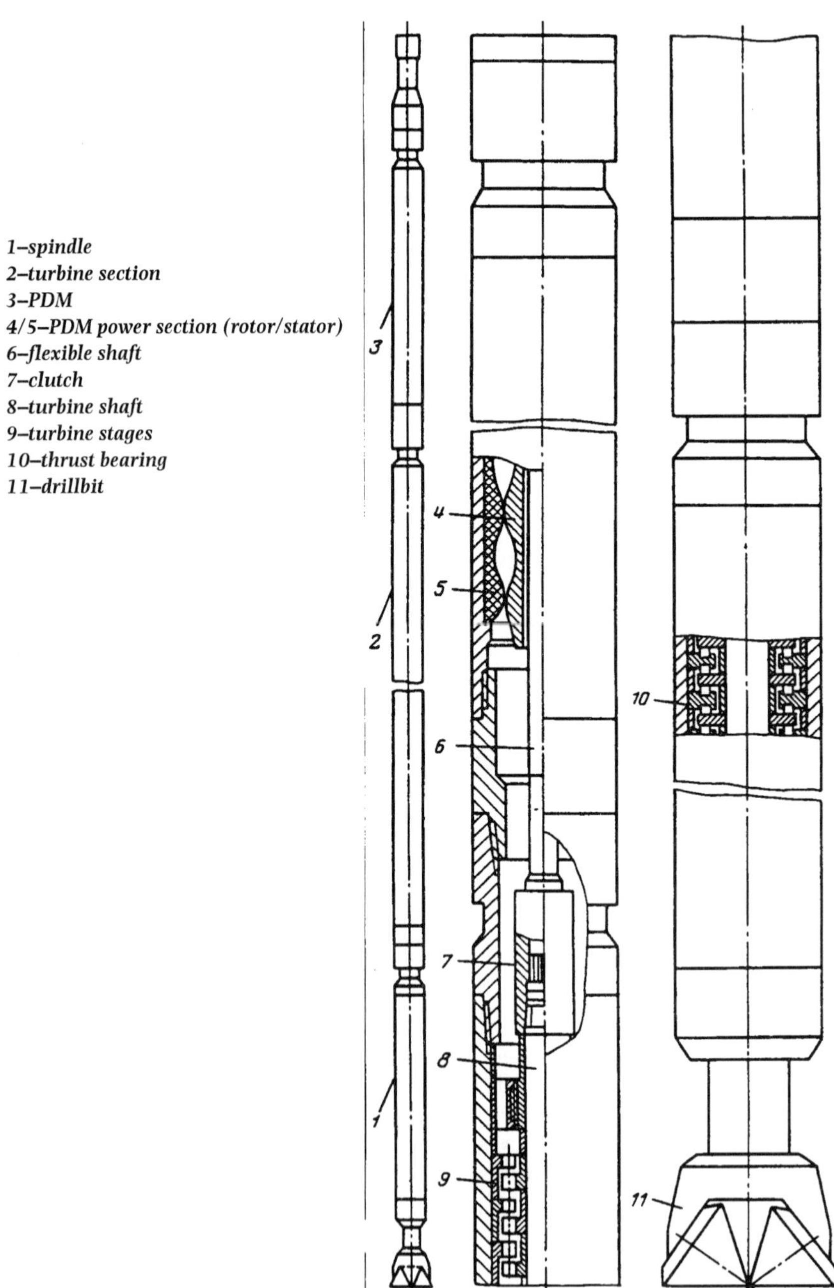

Fig. 2–46 Compound PDM plus turbodrill downhole motor

Obviously, the engineers were not able to change the characteristics of the non-geared turbodrills by using various types of turbines that were fabricated using conventional steel casting technologies. Turbodrill designers were searching for a solution by developing various designs, which enabled an increase in power characteristics per turbodrill unit length. This development work resulted in turbodrill designs that were commercially tested. They combined a high-speed high torque turbine with the hydro-mechanical or mechanical brake mechanisms (including mechanical transducers) and controlled the output characteristics, the extreme of which coincided with the zone of the optimum bit rotational speed. Another solution to the problem was the development of the multistage design concept with floating stators or rotors previously described.

Simultaneously, the engineers continued working on improving the turbine stage by reducing its height without affecting its power characteristics and performance results. These studies were carried out at the early phases of multistage turbodrills development but failed to find practical implementation in the designs of commercially manufactured turbodrills. Nevertheless, the increase in the number of stages as a natural way of improving the power characteristics of a turbodrill without gear-reduction could not be achieved without finding a solution to this problem. Reducing the turbine stage height was possible by reducing the stator and rotor blade cascade size, by reducing the turbine axial clearance, or both. The turbodrills with floating stators or rotors provided the best opportunity to do this.

In conventional turbodrill design, stage height could be decreased by reducing the cascade axial dimension blades. The risk of losing axial clearance and rotor interference with the stator is higher for a stage with smaller axial clearance. The objective was to determine a feasible method to decrease the blade cascade height, which would also ensure sufficient improvement in power of multistage turbines (standard three-section design) or to reduce the length of the turbodrill.

Given the geometrical similarity of the original and the reduced size blades, the power characteristics of a stage were preserved with few changes. However, in this case, the number of blades was higher, they were thicker, and the blade flow passage area was smaller. This complicated their operation in mud with high solids content and created problems when manufacturing the turbines. Therefore, during the development of these blade cascades, instead of the original blade profiles, those with a higher thickness ratio (c = c/b) were used at a higher pitch ration, compared to the optimum blade pitch ratio (t = t/b). Here, c represents the thickest blade profile, t represents the blade pitch in the average diameter zone, and *b* represents the blade profile chord. [36]

Naturally, a decrease of power was possible at such deviation compared to the original stage. Compensation for these losses might be achieved using special blade cascade designs. These methods achieved the following:

- the radial component, directed toward the turbine axis, was given to the stator outflow velocity
- the radial component, directed off the turbine axis, was given to the rotor outflow velocity
- the fluid loss through radial clearances was reduced
- the flow variation of blade passages from the blade root to blade head was reduced

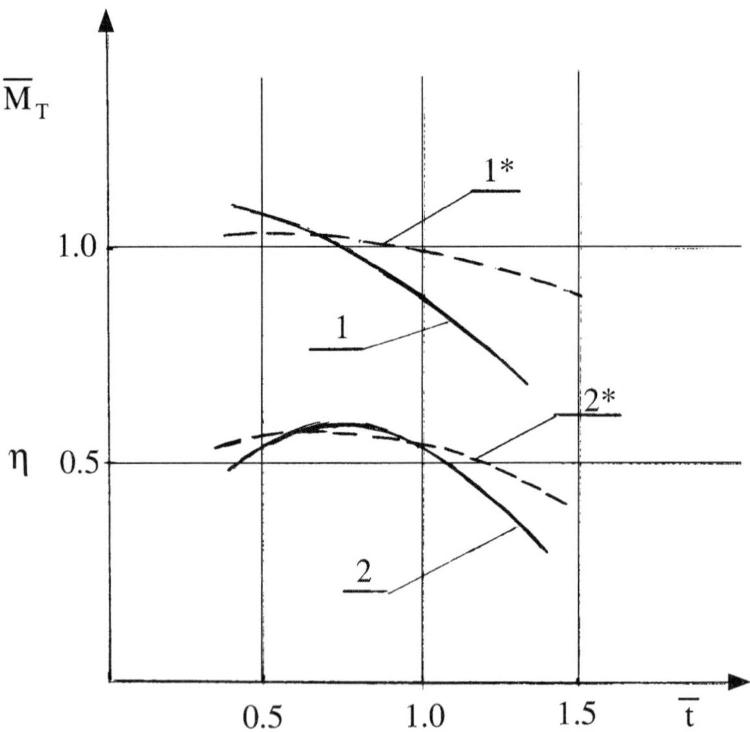

Fig. 2–47 Modern turbodrill torque (1) and efficiency (2) versus pitch-chord ratio of turbine stage: 1 and 2 for classical turbine design; 1* and 2* for special design development

On the whole, this achieved a more regular flow pattern in the flow passage along the turbine wheel radius and reduced the possibility of the flow "falling-through," i.e., reduced the deflecting ability at higher cascade pitch-chord ratios.

For such turbines, the high density and viscosity of drilling mud was a positive factor because it excluded the possibility of the flow "falling-through," and at a high cascade pitch-chord ratio, it helped reduce the need to slug the blade passage.

The diagram in Figure 2–47 shows experimentally built curves of the turbine stall torque M and its efficiency versus a cascade pitch-chord ratio t.

Curves (1) and (2) were built for $M(t)$ and $\eta(t)$ for the reduced blade cascade length based on the classical form of the geometrical similarity principle. Curves (1*) and (2*) were built in the same way for the new special blade designs. It is obvious from this diagram that in the second case, the increase of the cascade pitch-cord ratio t did not result in a substantial loss in the power characteristics of these turbines.

Considering the experience in designing reduced length turbines, the minimum acceptable blade cascade axial dimension for the low circulation degree (and propeller-type) turbines $h = 8...9$ mm. The blades cascade with $h = 9...12$ mm is for the high circulation degree turbine. This allowed building a standard design stage with height of $H = 30...40$ mm, preserving the conventional axial clearance of the turbine of up to 16 mm, or making it significantly smaller due to facilitated control of the reduced length turbodrill.

Use of the new turbines for both low-speed (with gear reducer) and high-speed turbodrills proved that this development trend was quite promising for turbodrilling applications.

Table 2–27 presents parameters of turbodrills 1T-240 and 1T-178 designed by the Aquatic Company (Russia) from 1996 to 1998 for TIEBO Tiefbohrservices (Germany) and Borais Petroleum Investments (Egypt) directional drilling companies. These designs were made in close cooperation with the customers to achieve the best performance and reduce the cost per meter to drill deep directional wells offshore and onshore. Some examples of these applications are presented in the next section.

TABLE 2–27
Modern Turbodrills Operational Characteristics
(Mud Weight 1.2kg/dm³)

Turbodrill code	Diameter (mm)	Length of spindle (m)	Length of turbine (m)	Flow rate (l/sec)	Pressure (MPa)	RPM at free rotation (min^{-1})	Stall torque (Nm)	Output power (kW)
1T-178S	178	2,8	10,2	24	6,0	1,320	2,000	70
				27	7,6	1,485	2,530	100
2T-178S	178	2,8	14,0	24	9,0	1,320	3,100	110
				25	9,8	1,375	3,360	125
1T-240S	240	2,6	10,8	40	7,5	1,200	4,800	150
				45	9,5	1,350	6,100	215
1T-240P	240	2,6	10,8	40	7,0	990	4,660	125
				45	8,9	1,115	5,900	180
2T-240P	240	2,6	18,2	36	9,7	890	6,610	140
				40	12,0	990	8,160	190

Modern turbodrill application outside Russia. Several foreign companies started studying the Russian turbodrills market in the early 1990s. Unfortunately, Russian turbodrills, especially the 9½-in., did not meet the requirements for deep directional well drilling in steerable mode. Finally, the prototype turbodrill design with K-type low height turbine stages was selected along with the Perm Branch VNIIBT geared turbodrill. These were the only tools with characteristics promising enough to compete with turbodrills available for worldwide steerable drilling.

The first two test runs of Russian 240-mm turbodrills were made by Borais Petroleum Investment Company offshore Egypt in 1996. The 12¼-in. PDC M73BP type drillbits were used in both runs.

Run 1. Turbodrill TO-240KE was similar to 1T-240 (*see* Table 2–27), except the section spindle connection. It was a flexible shaft instead of a U-joint. One-degree bent housing was used in this run. Cement plug in 13⅜-in. casing and 5079–6177 foot interval were drilled out. Total time consumed was 54.4 hours including 6.7 hours through the cement plug and 44 hours pure drilling. Average ROP evaluation was 28 foot per hour (fph), which was similar to competitive turbines.

Drilling practices were as follows: 660 gpm, 10,000–15,000 pounds-force (lbf) WOB, 70–80 rpm top drive, mud pressure 3600-4000 psi. The mud density increased from 8.9 to 13 ppg during the run. After building up the angle to 12°, the BHA was pulled out of the hole (POOH) because the pressure increased 35% more than calculated.

The bit showed significant wear on gage and peripheral cutters possibly caused by cement plug drilling. The turbodrill was disassembled onshore later, and the turbine was 10–15% plugged with cement slurry.

The manifold was cleaned out and a second run was conducted with a geared turbodrill and new bit.

Run 2. The TO-240R turbodrill consisted of a turbine section with the same type turbine and the TRSh-240 reduction gear spindle section. The connection was made with a conventional coupling, but the turbine had a thrust bearing suspension to reduce the axial load on the connection. A one-degree bent housing was used in this run.

The 6177–10,000 foot interval was drilled out. Total time consumed was 106 hours including reaming and washing. Average ROP was 36 fph, more during the steering operation and less when sliding. Sometimes ROP achieved 140 fph. The results were better than competitive turbines in the similar conditions.

Drilling practices were as follows: 660 gpm; 20,000–28,000 lbf WOB, 80–90 rpm top drive, 3600–4000 psi. Mud density was 13.5 ppg. After finishing the section, the BHA was POOH. The reducer was in good enough shape to be used longer.

Since 1997, 1T-240 type turbodrills were used extensively with PDC bits to achieve significant ROP improvement compared to PDM drilling. The 1T-172 design turbodrills were tested as well by the Borais Company in 1999-2000 with diamond impregnated bits.

The TIEBO Company, Germany, approached the Russian market aware of Russian drilling technology features, including turbo-drilling, because of scientific super-deep drilling experience. Specialists from TIEBO studied the latest results and trends in Russian turbodrilling R & D efforts with the help of the Aquatic Company and VNIIBT engineers in 1995 and 1996. TIEBO started testing Russian 240-mm turbines in 1997 in Germany and later in The Netherlands and Saudi Arabia. These turbodrills were equipped with U-joints to connect turbine and spindle shafts. After gaining experience, some modifications were implemented in 1999 to enhance turbine durability for applications in steerable and performance drilling technology.

One of the major tasks of TIEBO was the development of a $4\frac{3}{4}$-in. turbodrill for hard and hot rock applications. The new 120-mm design of geared turbodrills was heavily tested in different conditions during 1998 and 1999. Significant improvements were implemented, mainly in high-speed axial and radial bearings. Table 2–28 presents the results of some turbine runs made by the TIEBO Company from 1996 through 1999.

TABLE 2–28
Tiebo/Aquatic/Neftegaztechnika Turbodrills Field Examples

	RIH	POOH	Depth RIH (m)	Depth POOH (m)	Bend	Hours drilling	Mud weight (kg/cm³)	Flow rate, (l/min)	Pressure (bar)	RPM	WOB (t)	Avg. ROP (m/hr)	BIT	BIT size (inch)
1T240S		25.07.97	3,642	3,720	0.5°	52	1.67	2,100	285	100	16	1.50	GDK35X	$12\frac{1}{4}$
1T240P	08.08.97	17.08.97	3,835	4,124	0	159,5	1.69	2,000	285	110	16	1.81	DBSTBT601	$12\frac{1}{4}$
1T240S	02.09.97	11.09.97	4,131	4,520	0	172,5	1.69	2,200	285	120	13	2.55	DBSTBT601	$12\frac{1}{4}$
TR120P	29.12.97	02.01.98	3,130	3,258	0	57	1.2	850	150	40	4	2.25	S279G8	6
TR120S	11.04.98	13.04.98	5,206	5,303	0	30	1.42	800	240	100	6	3.2	CDPS279G	$5\frac{7}{8}$
1TR-240S	27.04.98	02.05.98	3,892	4,146	0	112,5	1.65	1,910-2,390	275	65	9–12.5	2.25	AG 547	12
1TR-240S	01.06.98	03.06.98	1,922	2,554	0	69,5	1.0	2,890	186	120	9–13.5	9.01	AG 437	12
1TR-240S	31.10.98	08.11.98	4,191	4,379	0	135,5	1.81	2,700	320	130	16	1.73	TBT 601	$12\frac{1}{4}$
1TR120P	18.11.98	21.11.98	3,011	3,065	0.75	54,5	1.55	620	186	70	4	1.0	Hy733XG	$5\frac{7}{8}$
1TR240S	27.01.99	29.01.99	2,008	2,540	0	44,5	1.0	2,839	158	90	11,3	11.9	AG 437	12
1TR240S	30.01.99	02.02.99	2,540	2,825	0	57	1.0	2,800	206	90	11,3	5.0	AG 437	12

Locations were in Germany, Netherlands, and Saudi Arabia.

In March 1999, the Hughes Christensen $12\frac{1}{4}$-in. S280 G2 used impregnated bit drills for a world record ROP with TIEBO turbodrill 1T-240S. This was the Clyde Petroleum well P/9-8 drilled from the Noble Lynda Bossler jack-up rig offshore The Netherlands. The registered ROP was 7.18 m/hr on average in 2880–3002 m intervals that consisted of Lower Cretaceous Vlieland sandstone.

The 172- to 178-mm turbodrills became the subject of tests in 2000 with diamond impregnated bits. Following is an example of multiple turbine use in one of the deep wells in Germany in the year 2000.

Deep Gas Well Verden Ost Z1, Wintershall, Germany, turbodrill experience. The well Verden Ost Z1 is a rotliegend sandstone gas well in the northern part of Germany. The total vertical depth (TVD) was 5100 m with a departure of 96 m to azimuth of 186°. The geological profile, the casing/liner depths, and the horizontal and vertical projection of the well are shown in the Figure 2–48.

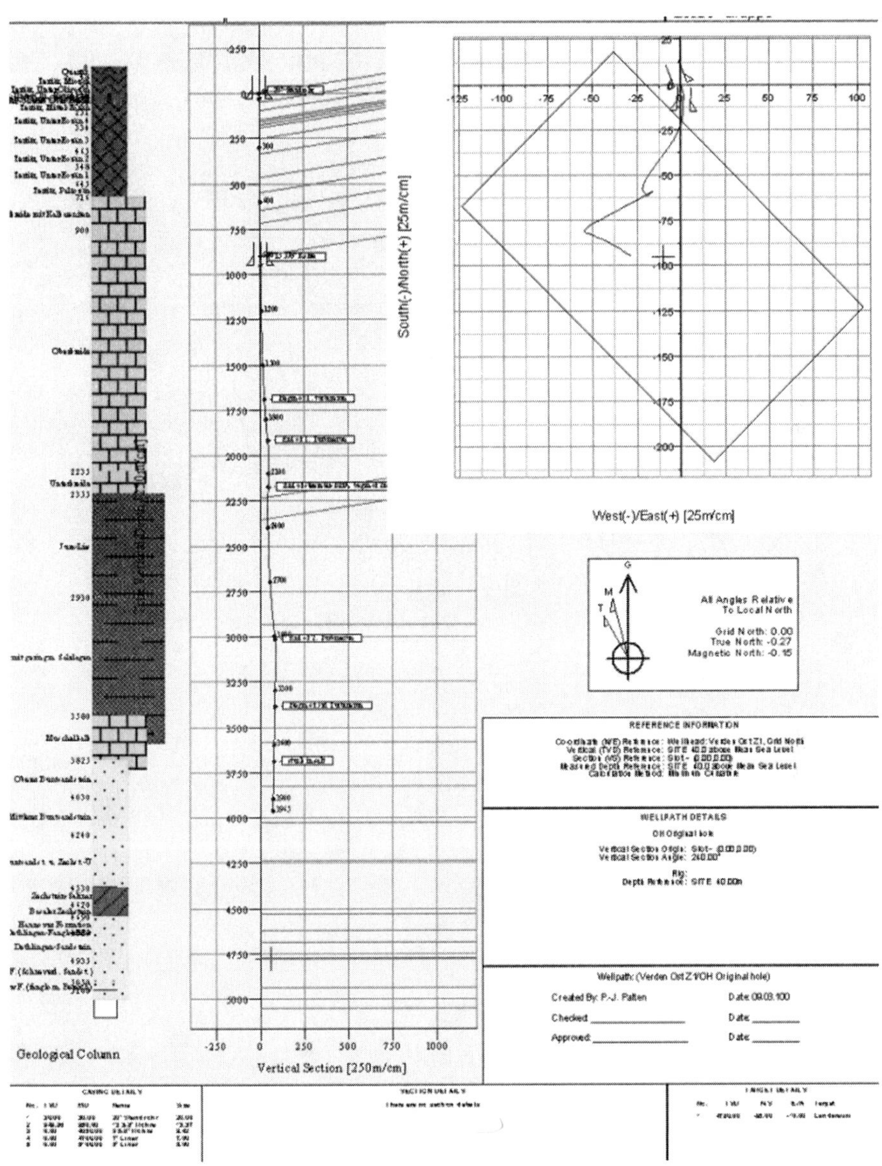

Fig. 2–48 The well Verden Ost Z1, Germany drilled with turbodrills application

The long $12\frac{1}{4}$-in. section was planned to be vertical and could be drilled in the upper section with a 400 to 800 rpm PDC bit. This is an ideal case for the $9\frac{5}{8}$-in. (1T-240) turbine without a gear reduction unit. In the lower part of the $12\frac{1}{4}$-in. hole section, the formations are harder so that the PDC needed more torque. This was the case for the $9\frac{1}{2}$-in. turbine with gear reducer (TR-240).

In the $8\frac{3}{8}$-in. hole section, hard and abrasive Buntsandstein rocks were drilled. The bit was an impregnated type driven by a $6\frac{3}{4}$-in. turbine without gear reduction (rpm 600–1000).

The $5\frac{7}{8}$-in. hole section composed of hard abrasive sandstone was successfully drilled with an impregnated bit (500–1000 rpm up to 1500 rpm depending on flow rate) and TR-120 geared turbodrill.

Table 2–29 presents details of turbodrill runs, and Figures 2–49a and 2–49b present the time/depth graph.

TABLE 2–29
Turbodrill Runs in Verden Ost Z1 Well, Germany in 2000

Turbine	Drillbit	Serial #	Depth RIH m	Depth POOH, m	Drilled m	Circ. Hours	Avg. ROP, m/h	Max. ROP, m/h	Flow-rate l/min	Mud density, kg/1	Reason for POOH
1T240S	PDC w/reamer SD65M	1,903,088	1,689.3	1,919	229.7	27	9.57	16.6	3,000	1.14	Build up -DD BHA
1T240S	PDC w/reamer SD65M	1,903,088	2,174	3,016	842	160	5.59	26.2	3,000	1.39	Bit worn
1T240S	PDC SD646	1,963,202	3,389.5	3,692	302.5	129	3.6	11.8	2,900	1.4	Bit got stuck in salt
1TR240	PDC bi-center SR144	97,300	3,791	3,965	174	52.5	9.16	21.4	2,500	1.6	Bit and turbine got stuck in salt
1T-178S	Impregn. S28 G8	1,211,397	4,085	4,505	420	194	2.21	5.2	1,600	1.46	Pressure increase 60 bar, bit worn
TR-120	Impregn. S279 G8	1,211,618	4,711	4,783	72	46.5	1.87	4.0	780	1.45	Bit worn
TR-120	Impregn. S279 G8	1,211,713	4,820	5,048.5	228.5	82	3.09	11	720	1.45	Finish section

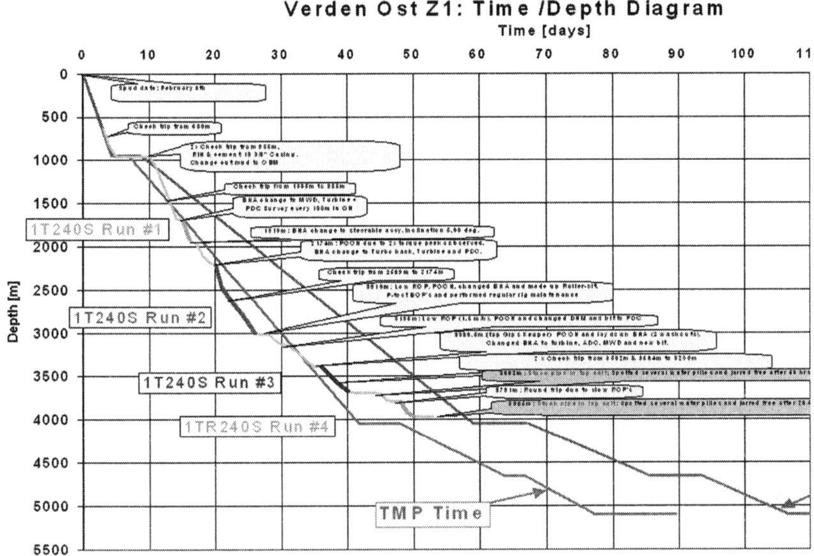

Fig. 2–49a The well Verden OstZ1 time-depth diagram, up to 4000 m depth

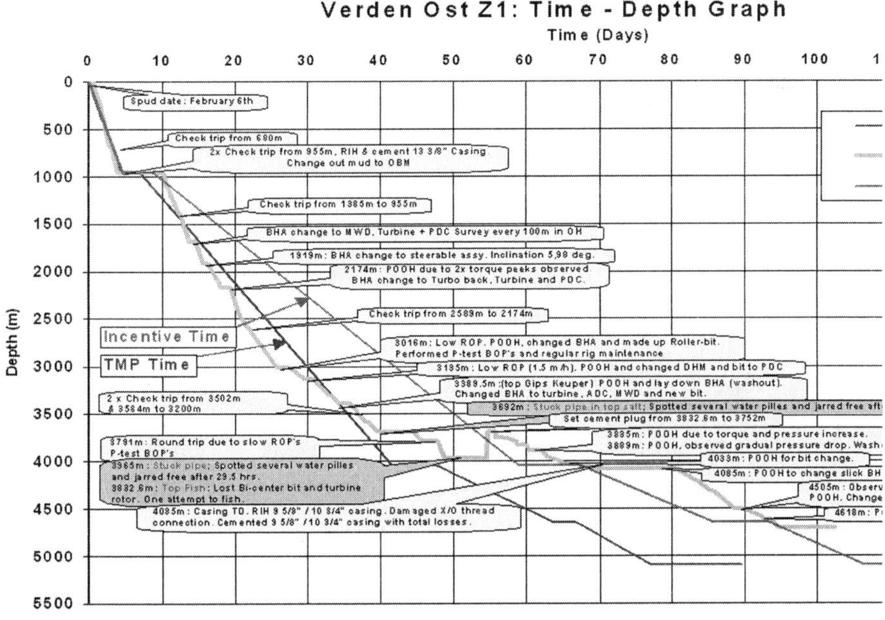

Fig. 2–49b The well Verden Ost Z1 time-depth diagram, up to 4700 m depth

The well was drilled with a 16-in. section by rotary drilling method from 0–950 m. A $13^3/_8$-in. casing was run and cemented to the surface. The mud system was changed to oil based with 1.20 g/cm³, and the $12^1/_4$-in. section was drilled rotary to 1690 m total measured depth (TMD) through flint stone. Then a fully stabilized BHA with a $9^1/_2$-in. 1T240S TIEBO Turbine was run with a SD65 pilot PDC bit (Figs. 2–50a, 2–50b, and 2–50c). The well was drilled to 1919 m TMD (229.70 m in 27 hours) through the chalk. The inclination was building continuously due to a 10–12° formation dip. The BHA was POOH to run a steering assembly with a motor plus TCI rock bit prior to drilling into the Lias formation. At 2174 m TMD, the BHA was POOH while the mud weight was increased to 1.4 g/cm3.

After the steering assembly run, a full-stabilized BHA with a 1T240S turbine (the same power section as in the first run, but a new spindle) was used. This turbine drilled from 2174 m TMD to 3016 m TMD with the same pilot bit. The total penetration was 842 m in 160 hours through the Lias/Keuper into the Räth sandstone. Here, the PDC pilot was completely worn Figure 2–50c.

The Räth sandstone was drilled with a PDM and a cone bit, code ATM22 to 3135 m TMD into Keuper-Steinmergelkeuper. The BHA was POOH because of low ROP and changed to a DPI PDC-bit SD65M with a PDM. This BHA drilled to 3389.5 m TMD and was POOH because of a washout in the box of a $12^1/_8$-in. string stabilizer.

The third turbine run was also with a DPI PDC low torque SD646 pilot bit. This bit is similar to the DPI SD65M equipped with 19-mm cutters but with smaller cutters. The bit drilled through Keuper, chalk, and upper Buntsandstein into salt before it got stuck. The drillbit was released after 58 hours of jarring; the flow rate was 2900 l/min with mud weight 1.4 kg/l.

Fig. 2–50a PDC bit SD65 used with 1T240 turbodrill in the well Verden Ost Z1, turbodrill assembled with turbo-back stabilizer and PDC bit—the special key is still on the turbine shaft above the stabilizer

Fig. 2–50b PDC bit SD65 used with 1T240 turbodrill in the well Verden Ost Z1, PDC pilot bit SD65 (new)

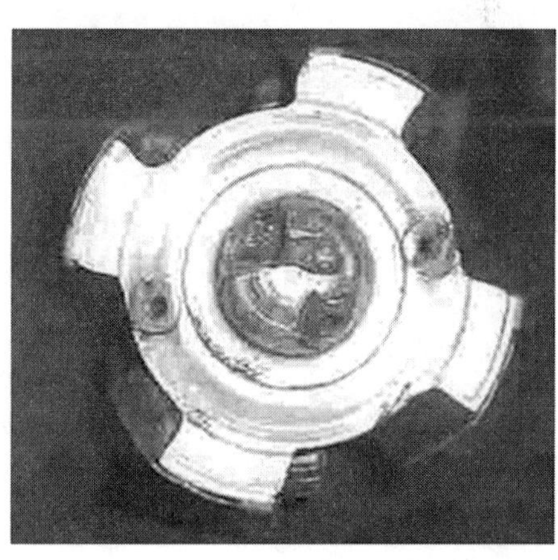

Fig. 2–50c PDC bit SD65 used with 1T240 turbodrill in the well Verden Ost Z1, SD65M bit serial # 1903088 after two turbodrill runs, total penetration 1071.7 m in 174.5 hours

Because of the problems in the salt, the next BHA was a rotary reaming BHA with an ATM M18 DGT and a roller reamer. This BHA drilled up to 3791 m TMD and was changed to a DPI-bicenter SR144 bit with a gear reduced turbine. The bit and turbine were stuck in the salt at 3965 m TMD. The BHA got free after jarring for 28.5 hours. The flow rate was 2500 l/min and the mud density 1.60 g/cm^3. The BHA was POOH but not to surface. The decision was made to run in the hole again. Because of intensive jarring and dynamic torque in the DS, the lowest turbine housing connection came unscrewed. The bit, turboback, and the lower part of the turbine housing were lost in the hole. After one try at fishing, the well was cemented and kicked off. The hole was drilled to 10$^3/_4$-in. casing depth at 4078 m TMD rotary.

Fig. 2–51a Turbodrill 1T-178S run with impregnated diamond bit, turbodrill assembly before run

Fig. 2–51b Turbodrill 1T-178S run with impregnated diamond bit, worn bit after 420 meters in 194 hours run

After drilling into the cement with rotary BHA $6^3/_4$-in. turbine, 1T178S was used together with an impregnated diamond bit and drilled 420 m before the bit was worn. Figures 2–51a and 2–51b show the 1T-178S turbodrill with bit and stabilizer before the run and the worn drillbit. There had never been a run through this Buntsandstein formation with one bit. The last 200 m to $7^5/_8$-in. casing depth were drilled with rotary techniques through the salt and sandstone.

In the $5^7/_8$" hole section, a 1TR120 gear-reduced $4^3/_4$" turbine was used with an impregnated bit of the same type as previously described (*see* Table 2–29).

The experience of the TIEBO and Borais companies shows the possibilities for the application of modern turbodrills in different borehole conditions, including heavy mud and deep wells. The 240-mm turbodrills had never been used in Russia at depths of more than 2000 m, but now we have good examples of successful applications of the 1T-240 turbodrill at depths of more than 4000 m. The $4^3/_4$-in. turbodrill should be efficient in high temperature deep wells including horizontal ones. That size turbodrill had never been used for the performance drilling. In general, there is a logical tendency to enhance turbodrilling operations because of the significant growth in the use of PDC and diamond bits. Published records indicate that more than 30% of the bits used worldwide in year 2000 were these types of bits, which had increased tenfold over the last decade.

PDM (Screw) DHMs

PDM principal design

As mentioned earlier, PDMs hold promise for various drilling applications. This subsection presents a description of various PDM designs, as well as the theory and principles of their operation, characteristics, and spheres of application.

From 1966 through 1975, designers and engineers carried out development work and built the commercially produced D-series PDM. Initially, engineers built and tested the motors with diameters of 85 mm and 127 mm. Experience indicates that PDMs perfectly match current drilling practices. They do not require use of special drilling tools, pumps, or other equipment. [37]

All PDMs had a uniform principal design. The D-type DHM (Fig. 2–52) consisted of two sections: a motor and a spindle.

The motor section included a stator (1), a rotor (2) that was essentially an eccentric screw mechanism, and a double-hinged joint (3). The stator was made as steel housing with a rubber faced inside surface that had 10 specially shaped spiral teeth. The stator was connected to a string of pipes through a sub. The steel rotor had one tooth less than the stator, and its axis was offset from the stator axis by an eccentric value. The two-step working elements (stator and rotor) were used in the motor.

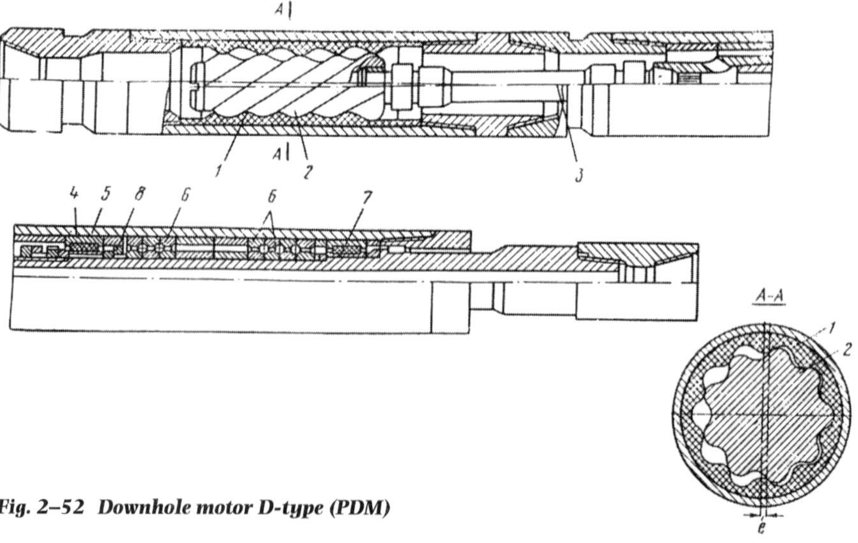

Fig. 2–52 Downhole motor D-type (PDM)

The screw surfaces profile could be based on the centroid and off-centroid hypo- and epicycloidal engagements. The rotor and stator screw surfaces were left-handed to enable clockwise rotation of the rotor (relative to its axis). Through a double-hinged joint (3), torque was transmitted from the rotor to the output shaft (4) installed in the spindle section. Besides the shaft, the spindle section contained a housing (5), multi-row axial bearing (6), radial rubber-metal bearing (7), and a seal (8).

Similar to the turbodrill spindle, the PDM spindle transmitted the axial load to a bit, takes up the hydraulic load, applied it to the motor's rotor, and sealed the shaft lower end. This helped create the required differential pressure at the bit. Also, bearings took up the radial loads generated by the drive shaft.

PDM working cycle

Although the screw motors had a simple design, their working cycle was quite involved. The main principles are described briefly as follows. [38] This issue was described in more detail in the book by D. F. Baldenko, F. D. Baldenko, and A. N. Gnoevykh. [39]

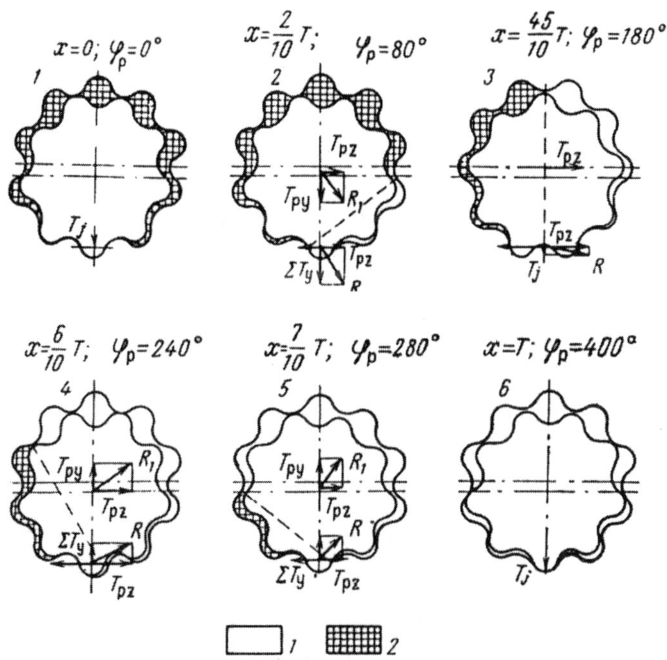

Fig. 2–53 Pressure distribution in the multi-lobe PDM working elements hollows—the cross sections along the X-axis on the T-pitch length; rotating angle of the rotor screw surface: 1–high pressure; 2–low pressure

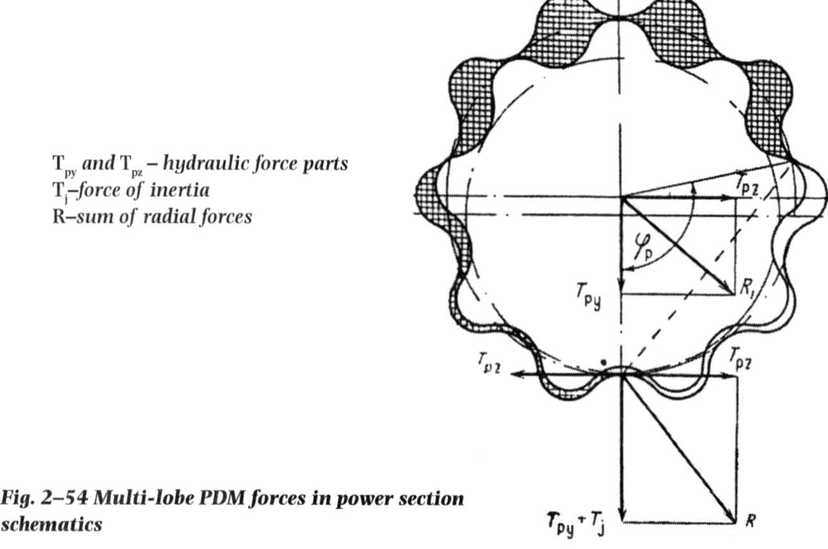

Fig. 2–54 Multi-lobe PDM forces in power section schematics

The rotor and stator screw surfaces divided the power volume into several hollows. The fluid chambers, Z_1 , existed in any cross-section and divided by contact lines (Fig. 2–53). During the operation of the motor, each of the Z_1 chambers periodically communicated with the high- and low-pressure hollows. Theoretically, the separation of the hollows with fluid, located above and below the motor's working elements, occurred at the distance equal to the stator pitch length. The unbalanced force R_1, applied to the rotor, is created in each transverse section of the working elements at the rotor pitch length (Fig. 2–54).

When a second cross-section is at a distance of *dx* from the one in question, the hydraulic force *dR1*, applied to the singled out element is

$$dR_1 = p\Pi dx \qquad 2.30$$

where

p is the differential pressure in the motor

Π is the length of an area to which the hydraulic force is applied

The analysis of the hydraulic force effect in the rotor and stator screw surface sections from φ_p = 0–40 indicated that normally the area began from the engagement line—in the point of contact between the two circles. For a multiple thread screw mechanism, the area length is accurately determined using the equation

$$\Pi = D\sin\frac{\varphi_p}{2} \qquad 2.31$$

where

D is the designed diameter of the eccentric rotor mechanism

$$D = 2eZ_1 \qquad 2.32$$

In the previous and following equations, the variable *e* is the eccentricity of the eccentric mechanism.

Specialists from the VNIIBT carried out a theoretical study and derived the following formula for torque calculation:

$$M = M_0 pDet \qquad 2.33$$

where

M_0 is the specific torque and is found from the equation

$$M_0 = \frac{Z_2 - 1}{2} + \frac{2}{\pi c_e} \qquad 2.34$$

Hence, the specific torque M_0 is a function of the rotor settings number Z_2 and a non-dimensional parameter C_e, found from a proportion between e and toothing radius *r*.

The physical meaning of specific torque stems from Equation 2.33. It is a torque of the eccentric mechanism with unit dimensions (*D*, *e*, and *t*) and unit pressure.

For a mechanism with a single start rotor ($Z_2 = 1$), this parameter has a minimum value ($M_0 = 3.16$ at $C_e = 0.2$). Increase of the rotor settings number results in higher M_0.

Using values of the designed diameter *D* and the non-dimensional parameter C_t (where $C_t = t/D$) for the Equation 2.33, the following formula is derived:

$$M = M_0 \frac{pC_1 D^3}{2Z_1} \qquad 2.35$$

An analysis of factors affecting screw motor rotational speed n is presented as follows. The following equation is true for all volumetric motors including screw motors:

$$n = \frac{Q}{V_0} \qquad 2.36$$

where

Q is the flow rate

V_0 is the volumetric power of the motor (fluid volume per rotation)

For the eccentric rotor mechanism,

$$V_0 = F_{ch} T Z_2 \qquad 2.33$$

Since a kinematic pattern determines the proportion between the transportation and the relative motions as

$$\frac{\omega_{Tr}}{\omega_{Rel}} = Z^2 \qquad 2.38$$

In the eccentric rotor mechanisms, designed based on the hypocycloidal centroid engagement, the effective cross-section of the work element (chamber area) is calculated using the following formula:

$$F_{ch} = 2\pi e2(Z_2\text{-}1)+8er \qquad 2.39$$

Certain substitutions and transformations of Equation 2.36 allow deriving the formula for calculation of the screw motor output shaft rotational speed as

$$n = n_0 \frac{Q}{e^2 T} \qquad 2.40$$

where

n_0 is the output shaft specific rotational speed, calculated using the equation:

$$n_0 = \frac{1}{[2\pi(Z_2 - 1) + \frac{8}{C_e}]Z_2} \qquad 2.41$$

Like M_0, n_0 is a non-dimensional parameter and is a function of the mechanism's number of starts (Z_2) and the C_e coefficient. The physical meaning of this value is the rotational speed of an eccentric rotor mechanism with unit dimensions and unit flow rate.

Single start or single lobe rotor mechanisms—high-speed motors—have the highest n_0 value ($n_0 = 0.25$).

Research engineers from VNIIBT carried out a special analytical study to compare the operational characteristics of downhole screw motors at various kinematic ratios and their working elements. At the first stage of the study, the researchers analyzed geometrically similar eccentric rotor mechanisms with constant C_e and C_t parameters, assuming constant differential pressure and flow rate.

In particular, they calculated output parameters of the motors with a diameter $D_{\delta\varepsilon}$ = 172 mm. The flow rate was Q = 25 liters/sec, where the differential pressure for one pitch, Δp = 10 kg/cm². Figure 2–55 presents the results obtained for motors with similar geometry. The results of these studies were first published by Gusman and D. F. Baldenko in the article. [40] Their analysis was further developed in the book. [41] These results demonstrate the principal advantages of the multi-lobe screw motors. In the following years, many western companies used similar diagrams as impressive graphics for multi-lobe PDM presentations.

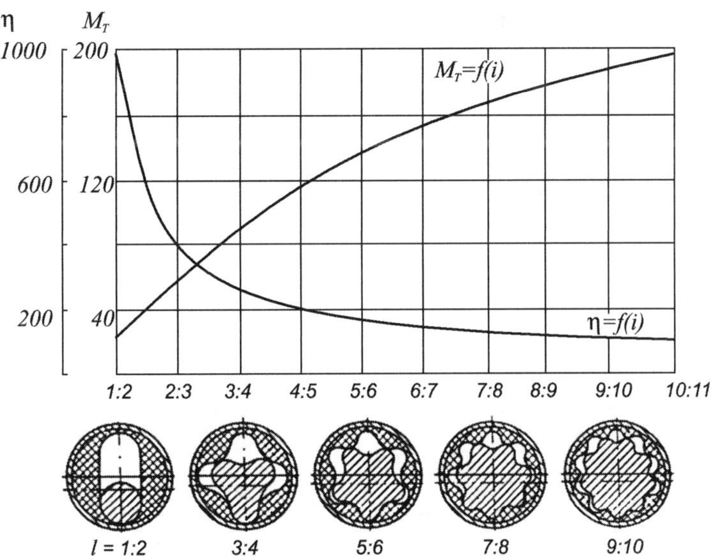

Fig. 2–55 Diagram of PDM characteristic dependency on the mechanism kinematics, number of lobes

Accordingly, motors with eccentric rotor mechanisms and few starts achieved high rotational speed and low torque. An increase in the number of rotor starts (lobes) resulted in an increased torque and a decline in rotational speed because the multiple thread eccentric rotor mechanism was essentially a combination of a hydraulic motor and a planetary reduction gear with a gear ratio proportional to the number of rotor starts. This is extremely important for the output parameters and the motor load-carrying capacity.

The contact lines of the working elements in volumetric motors were quite long. During DHM operation, when abrasive fluids were pumped along the contact lines, fluid started leaking out of the working chambers.

As in all volumetric motors, the output shaft rotational speed theoretically is not dependent on torque. In reality, screw motors feature a relatively flexible parameter $n = f(M)$, compared to other volumetric motors, due to variations in rotor alignment.

Variable rotor alignment means that the elastic facing of the stator becomes radially deformed, affected by hydraulic and inertial forces, and redistributes the initial tension between the rotor and the stator. The resulting gap allows fluid to flow from the working elements' chambers toward the low pressure hollow.

The volume of the fluid leak, q_{yT}, for the definite design of the working elements, hinges on the differential pressure level, the tension between the rotor and the stator, the viscosity of the pumped fluid, and the hardness and thickness of the stator elastic facing

$$q_{yT} = \mu Q_{cp} L_Q \sqrt{\frac{2gp}{\gamma_f [2Z-1]}} \qquad 2.42$$

where

μ is the flow rate coefficient

Q_m is the average size of clearance between a rotor and a stator

L_Q is the length of the clearance

γ_f is the fluid specific gravity

Z is the number of working element stages ($Z = L/T$)

L is the length of the working elements

Using Equation 2.42, the analytical dependence of the volumetric efficiency factor from operating conditions can be determined using the formula:

$$\eta_0 \approx 1 - \frac{S\sqrt{p}}{Vn} \qquad 2.43$$

where

S is the clearance area outlined by the contact lines of the working elements

V is the operational volume of the motor

Equation 2.43 indicates that the differential pressure increase results in a decrease in the volumetric efficiency factor proportional to the decline in motor torque. In contrast, rotational speed increase causes growth in the efficiency factor.

Characteristic flexibility and output power of a volumetric motor are dependent, to a large extent, on the stability of the volumetric efficiency factor. Therefore, motors with a rotor usually feature high rotational speed.

Low-speed high-torque PDM application

Under the conditions in question, when low-speed motors were used to achieve the optimum roller-cone bit operation, the only way to maintain the level of the volumetric efficiency factor was by an increase in effective power. However, an increase in the effective power caused a linear increase in the clearance area, which prevented achievement of desirable results.

An increase in the number of starts (*see* Fig. 2–56) caused an abrupt reduction in the length of the working elements and in the required eccentricity level (which, in turn, limited the service life of the rotor-to-output shaft joint assembly).

Using the multiple thread eccentric rotor mechanisms, we can design a range of high-torque motors with acceptable dimensions that are capable of operating at rotational speeds of 100–200 rpm or less and have high load-carrying capacity.

The design of the eccentric screw rotor mechanism has potential with regard to variation of the DHM output parameters. By changing the rotor-stator gear ratio, it is possible to vary the degree of torque reduction and rotational speed of the output shaft.

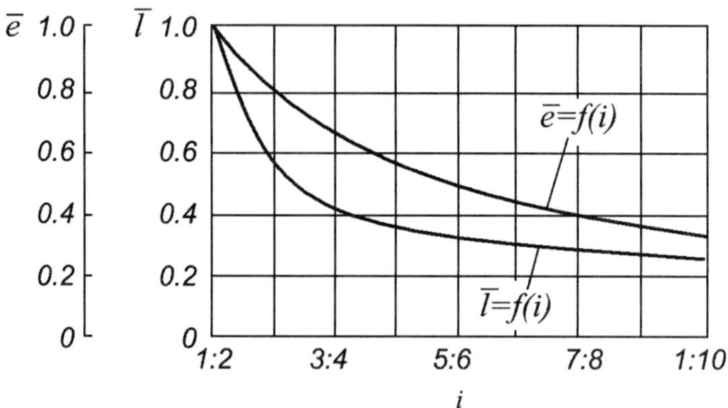

Fig. 2–56 PDM design parameters dependency on rotor/stator lobes ratio: e– eccentricity; l – rotor-stator length

An advantage of the multiple-thread screw, or a multi-lobe motor, is less dependence between the torque and the diameter of the motor. A 172-mm diameter motor has higher torque compared to the same diameter turbodrill. This enables drilling with 214-mm and 243-mm diameter bits and maintains higher clearance between the motor and the borehole wall. One important issue is the improved durability of the motor's working element, which is achieved by enhancing the motor design, increasing rotor strength, and using wear- and heat-resistant rubber facing elements for stators.

In 1974 the industry began producing two motors of standard type and size, D2-172M and D-85, and in 1976, the D1-54 was produced. The batch production of motors became possible thanks to a large volume of scientific R&D work carried out by specialists from the VNIIBT and its branch in Perm. Initially, large amounts of experience were accumulated during field testing of the D2-172M and D-85. In the late 1990s, screw-type DHMs found the following applications in Russia:

- standard well drilling
- directional and horizontal drilling
- well workover

Tables 2–30, 2–31, and 2–32 show characteristics of the motors in commercial use with rotary and turbodrilling methods by drilling companies in Russia today.

From 1969 through 1997, the motors made 350,000 runs in combination with roller-cone, blade, and diamond bits with diameters from 190 mm to 269 mm in depth intervals from 150 m to 5560 m. More than 26,000,000 m were drilled mostly in Tatariya, Bashkiriya, the Perm region, in Azerbaijan and Kazakhstan. [42] By 1997, the annual penetration volume of PDMs was 10% of the total footage drilled in the country (Table 2–33).

TABLE 2–30
General Duty PDM Parameters

Code	OD, mm	Length, mm	Mass, kg	Cinematic ratio	Active stator part length, mm	Recommended bit diameter, mm	Flow rate, l/sec	RPM	Pressure drop, MPa	Torque, kNm
D1-127	127	5,800	420	9:10	2,000	139.7–158.7	15-20	200–250	5.5–8.5	2.2–3.0
D1-145	145	4,670	418	7:8	1,800	158.7–190.5	15-20	120–180	7.0–9.0	3.0–4.5
D-155	155	4,870	500	7:8	2,100	190.5–215.9	24-30	125–160	6.5–7.5	3.5–4.0
DM-172	172	12,000	1300	7:8	6,300	190.5–215.9	18-32	180–200	8.0–10.0	6.0–7.0
DM1-172	172	12,540	1200	7:8	5,700	190.5–215.9	20-25	110–135	10.0–12.0	6.0–7.0
DN-172	172	3,900	490	9:10	1,800	190.5–215.9	25-35	80–110	4.5–7.0	4.5–6.0
D_-172	172	6,800	912	9:10	1,800	190.5–215.9	25-35	80–110	3.9–4.9	3.1–3.7
D5-172	172	6,220	770	9:10	1,800	190.5–215.9	25-35	80–110	4.5–7.0	4.5–6.0
D5-172_	172	6,720	830	9:10	2,400	190.5–215.9	25-35	90–120	7.2–9.7	7.4–9.8
DV-172	172	5,420	650	6:7	2,400	190.5–215.9	24-32	180–220	6.2–9.2	5.0–6.5
D2-195	195	6,550	1100	9:10	1,800	215.9–244.5	25-35	90–115	4.3–6.7	5.2–7.0
D5-195	195	7,265	1200	9:10	2,400	215.9–244.5	25-35	80–110	7.9–9.9	9.5–11.0
D3-195	195	7,940	1300	9:10	3,000	215.9–244.5	25-35	80–130	9.2–11.2	10.5–13.7
D1-240	240	7,570	1746	7:8	3,000	269.9–295.3	30-50	70–130	6.0–8.0	10.0–14.0

TABLE 2–31
Parameters of PDM for Directional and Horizontal Drilling

Code	OD, mm	Length, mm	Mass, kg	Cinematic ratio	Active stator part length, mm	Recommended bit diameter, mm	Flow rate, l/sec	RPM	Pressure drop, MPa	Torque, kNm	Maximum bent grad
DG-60	60	2,300	34	5:6	755	76.0–98.4	1–2	180–360	4.5–5.5	0.06–0.08	1,5
DO-88	88	3,570	85	5:6	1,080	112.0–120.6	5–7	180–300	5.8–7.0	0.4–0.6	5
DG-95	95	2,640	41	6:7	1,420	112.0–139.7	6–10	120–200	4.5–6.0	0.6–0.9	4,0
DGU-95	95	4,500	150	6:7	1,420	112.0–139.7	6–10	120–180	4.0–5.0	0.6–0.9	4,0
DG-105	106	2,355	120	6:7	1,000	120.6–139.7	6–10	170–240	5.0–8.0	0.6–1.0	5
DG-108	108	2,565	147	7:8	1,400	120.6–151.0	6–12	80–150	3.5–5.5	0.8–1.3	4
DG-155	155	4,330	466	7:8	2,100	190.5–215.9	24–30	125–160	6.5–7.5	3.5–4.0	3,5
DG1-172	172	3,870	455	6:7	1,460	190.5–215.9	24–35	150–190	5.8–7.8	3.5–4.0	3

TABLE 2–32
PDM for Workover Operations

Code	OD, mm	Length, mm	Mass, kg	Cinematic ratio	Active stator part length, mm	Recommended bit diameter, mm	Flow rate, l/sec	RPM	Pressure drop, MPa	Torque, kNm
D-35	35	1,600	15	4:5	560	Special bit	0.8–1.0	450	4.0–5.0	0.02–0.025
D-48	48	1,850	18	7:8	685	59–76	1.2–2.6	245–400	4.0–5.0	0.08–0.1
D1-54	54	1,890	27	5:6	530	59–76	1.0–2.5	180–450	4.5–5.5	0.07–0.11
D-85	88	3,230	110	9:10	870	98.4–120.6	5.0–7.0	225–290	4.0–5.0	0.5–0.6
D1-88	88	3,225	110	5:6	1,220	98.4–120.6	4.5–7.0	160–300	5.8–7.0	0.8–0.095
D-95	95	3,000	90	6:7	1,420	112.0–139.7	6–10	120–180	4.0–5.0	0.6–0.9
D1-105	106	3,740	180	5:6	1,500	120.6–151.0	6–10	155–230	5.0–8.0	0.8–1.4
D-106	106	4,200	215	7:8	2,000	120.6–151.0	4–12	35–125	4.0–8.0	0.9–2.0
D1-106	106	5,255	275	7:8	2,000	120.6–151.0	4–12	35–125	4.0–8.0	0.9–2.0
D-108	108	2,900	150	7:8	1,400	120.6–151.0	6–12	80–150	3.5–5.5	0.8–1.3
DK-108-1	108	5,000	250	14:15	1,400	120.6–139.7	3–6	20–40	5.5–7.5	2.0–2.7
D_-108-2	108	3,000	150	7:8	1,400	120.6–139.7	6–12	80–150	3.5–5.5	0.8–1.3
D_-108-3	108	3,000	150	4:5	1,400	120.6–139.7	6–12	120–240	3.0–5.0	0.5–0.8
D1-108	108	3,110	170	7:8	1,400	120.6–151	6–12	80–150	3.5–5.5	0.8–1.3
D-110	110	4,200	235	9:10	2,000	120.6–151.0	4–12	25–100	3.0–7.5	1.1–2.3
D-120	120	3,735	245	7:8	2,000	139.7–151.0	8–12	90–125	6.0–9.0	1.5–2.5

TABLE 2–33
PDM Application in FSU (Until 1990)
and Russia (1991–1997)

	1976–1980	1981–1985	1986–1990	1991–1995	1996–1997
PDM drilling volume, million meters	0.4	5.3	12.6	6.5	1.75
PDM drilling volume in percents of total oil and gas wells drilling in the country	0.6	4.2	6.6	6.9	10.0

PDMs were successfully used for deep vertical and directional drilling applications, as well as for drilling with coring using the Nedra core barrel (Russian analogue to Christensen 250P system).

The D2-172 motor successfully made 40 runs at the depth from 4332 m to 9040 m in the Kola Super-deep SG-3 borehole (*see* Volume 2, Chapter 5).

The experience gained in using volumetric motors made it possible to single out the following areas of efficient application:

- drilling lower well intervals when turbodrill utilization fails to be economical because of the low durability of tri-cone bits, i.e., when the optimum rotational speed should be less than 200–250 rpm
- drilling using heavy mud with a specific gravity of 2.0 g/cm^3 and higher
- drilling directional and horizontal well applications with severe inclinations in the angle of the build section using a small diameter and short length motor
- sidetracking from shut-in wells
- using the motor for well workover operations, such as cement drill out in production casing

The case study examples that follow illustrate the effects of using the screw motors for these applications. [43]

PDM case studies

In Tatariya, a typical geological section is primarily composed of hard rock confined to Paleozoic, from Upper Carboniferous to Frasnian deposits. The motors were used in combination with Russian sealed and non-sealed bearing bits and Western sealed bearing bits. The wells average depth was 1700 m, and drilling intervals below the conductor casing from 1350–1400 m. The results are presented in Table 2–34.

TABLE 2–34
PDM Application in Tatariya

Bit type	Open bearing 1V-190T, 1V-190K			Sealed bearing, home made 2AN-215.9	Journal bearing USA 8$\frac{1}{2}$"
Bit drive	rotor	Turbodrill 3TS5.B-168		PDM D2-172N	
Average penetration per bit (m)	70	30	60	175.8	274.8
Average ROP, m/hr	7.5	20.0	10	14.7	15.7
Number of bits	20	46.6	23.3	8	5

Note: Open bearing bits were 190-mm diameter against the 215.9-mm for sealed bits. Nevertheless, the comparison was very informative.

Use of an improved design D1-195 motor with a 195-mm diameter achieved even better results. Drilling two wells in similar geological sections using the motor with the M-88 and 8 ½-in. diameter drillbits required only 2.5 bits per well. Later, this type of motor was widely used in drilling. Table 2–35 presents a comparison of bench test data for characteristics of the D2-172M and D1-195 motors at water flow rates of 25–35 liters/sec. [44]

TABLE 2–35
PDM of 172 and 195 mm OD Parameters Comparison

Operating mode	D1-195	D2-172M
Idle:		
RPM	100–140	150–210
Pressure drop (MPa)	1.7–2.8	1.5–2.5
Maximal power:		
RPM	80–110	120–150
Torque (N m)	3,200–3,800	3,000–3,600
Power (Kw)	26–43	37–55
Stall:		
Torque (N m)	6,000–7,000	4,000–4,800
Pressure drop (MPa)	9–10	7.5–8.5

Table 2–36 compares results of the wells drilled by Almetyevsk Drilling Department using 215.9-mm diameter bits with a different type of drive. The TKZTsV-215.9 Russian-made open-bearing bits were used here only for turbodrilling, whereas the Russian GNU-215.9 and U.S.-made M-88-8½ bits with sealed bearings were used with other types of drives. The information in Table 2–36 indicates that the worst results for penetration rates regarding trip time (bit run speed) were shown in rotary drilling, whereas the best results were recorded when using the D1-195 with U.S.-made bits.

TABLE 2–36
215.9 mm (8½") Drillbits Performance with Different Types of Drive

Downhole Motors	Drilling interval (m)	Penetration per run (m)	ROP (m/hr)	Bit run speed (m/hr)
Unified turbodrill TSSh1-195	200–1,700	57.9	22.9	10.1
PDM regular D2-172M + sealed bearing bit of GNU type	600–1,700	105	13.2	9.1
PDM D1-195 + bit GNU	300–1,700	147.3	14.9	11.0
PDM D1-195 + USA-made bits	404–1,716	404	17.9	14.1
Rotor (80 RPM) + bits of GNU type	213–1,600	154	11.2	7.3

Most drilling during the following years was carried out using homemade bits. Still, the drilling results using motors with sealed bearing bits were always better, compared to non-sealed bearing bits. However, in the Urals and the Volga region, drillers used turbodrills to drill upper well intervals from 1000 m to 2000–2500 m, and PDMs for drilling the lower intervals. This combination helped achieve the highest drillbit run speed for the entire well interval (see Chapter 3 for parameter details). The results of D2-172M motor tests, performed while drilling deep wells offshore Azerbaijan on the Caspian Sea shelf, were quite positive. [45]

In 1975 and 1976, drilling companies used the motors to drill intervals from 4028 m to 4965 m in offshore well Nos. 20 and 26 in the Bulla-More field. Totals of 1411 m were drilled in both wells during 1202 hours. Total operating time for the motors, taking into account borehole reaming and conditioning, was 1552 hours. Although the specific gravity of the mud was 1.9–2.1 g/cm^3, the motor MTBF was 99.6 hrs. In addition, torque backup made it possible to use of 269-mm bits.

Table 2–37 presents a comparison of drilling results achieved while using screw motors, turbodrills, and rotary drives in similar geological conditions. The D2-172M motor with diamond bits was tested on other wells in this field. The test results proved the feasibility of using the motor when drilling with high density mud, when the complications mentioned previously prevented the use of turbodrills. Compared to rotary drilling, it shows better performance results thanks to higher bit rotational speed and increased penetration rate. Yet, while drilling with bits of 269-mm diameter, the results can only be achieved by using the advanced design, high-powered D1-195 motors that feature a spindle shaft seal to enable jet bits operation. During the following years, this BHA found practical applications.

TABLE 2–37
Comparison of Drilling Results
Offshore Azerbaijan, Bulla-More Field

	Borehole No.			
	20 D2-172M	21 Turbodrill	26 D2-172M	28 Rotor
Drilling interval (m)	4,028–4,965	4,196–4,976	3,364–4,182	3,309–4,379
Total penetration (m)	914	690	497	623
Penetration per bit (m)	23.4	14.4	33.1	27.8
ROP (m/hr)	1.05	1.62	1.5	1.05
Average bit run speed: penetration / (drilling + tripping time) (m/h)	0.42	0.32	0.6	0.46

Flow rate 22–28 l/sec; mud weight 1.75–2.15: open bearing 269.9mm drillbits.

The results of the PDM tests carried out by the Neftekumsk Drilling Division in the Stavropol region were quite noteworthy. These tests were important for two reasons: high downhole temperature typical of the regional geology and the use of oil-based mud. Therefore, these tests allowed drillers to obtain valuable information about the performance of the motors in these conditions. During the tests, 5 motors were used to drill intervals in 10 wells. The total penetration was 4924 m in 834.5 hours in the interval from 1264 m to 3625 m. Total motor operating time, including borehole reaming and conditioning, was 1009 hours. In the lower sections of the wells, the bottomhole temperatures were from 113° C to 142° C at non-circulating drill mud regime, where mud temperatures at the surface were from 50° C to 70° C. The drilling mud contained 1–4% cuttings and up to 9% oil.

The results of the tests indicated that thanks to use of the PDM, the average footage per bit run increased by 14%, and penetration rate and drilling speed increased correspondingly by 37% and 27%. These results agree with the results of the studies carried out by drillers from Stavropolneftegas Company when they drilled key technological wells (*see* Chapter 3).

The studies revealed that the optimum bit rotational speed for the larger part of the well interval was 120–150 rpm, which corresponded to the motor characteristics. This proved for the first time in the Stavropol region that drilling with the PDM could be more efficient than rotary drilling in some well intervals. [46]

The MTBF of the motor was 71 hours and was determined mainly by the rotor and stator durability. Therefore, research engineers carried out special work to improve the stator durability. Four types of rubber and polyurethane were used to make the stator's elastic liner. Stators showed the best results with an elastic liner made of IRP-1226 nitrile rubber. Seven rotors were used with this type stator during 495 hours of the stator service life. Stators with liners from polyurethane and other materials did not show positive performance results.

One method to extend the total motor operating period at normal conditions used by the engineers was to increase the diameters of replaceable rotors that were installed in sequence inside the stator. This compensated for stator wear, maintained the required clearance between the rotor and the stator, and maintained a longer period of needed power characteristics without increasing the mud circulation rate.

In addition to these examples of successful PDM use, drilling companies in the FSU achieved good results using these motors in regions such as Bashkiriya, Kazakhstan, Western Siberia, and Perm.

The PDMs built by Russian engineers won worldwide recognition in the drilling industry. This acceptance was proved when the Drilex Company bought several licenses for manufacturing Russian-designed motors, which later became top-ranked tools in the drilling industry.

Sectional PDM

The results of PDM operation in FSU regions where medium- and heavyweight drilling mud was used for bottomhole cleaning showed that the working elements and other motor assemblies lacked durability, which affected their economics. Further improvements in the motor design were introduced to the second-generation motor design, such as multi-pitch pattern utilization, strengthening the rotor work surface, quality improvement of stator rubber lining and attachment to the housing.

However, all these improvements were insufficient to significantly advance PDM technology, which prompted research engineers to begin work toward the design and construction of sectional motors. Work focused on the 195-mm diameter, which was used most often. The research work indicated that the increased length of the working elements, i.e., the multistage pattern, had a positive effect on the power and durability of the motors. Yet, the development of a single multistage motor faced as many difficulties as constructing a multistage turbodrill.

The engineers suggested a practical solution that sectioned the PDM, similar to the method that had been used before. They also drew on the experience gained from the development of sectional turbodrills when they expanded the new designs. After several years of continuous R & D work, the engineers built sectional PDMs that were successfully tested and that found wide drilling applications. [47]

Figure 2–57 shows the D1-195 motor design that consists of three assemblies: a motor section (2), a spindle section (3), and a relief valve (1). Figure 2–58 shows a sectional PDM with rotors connected through hinged joints. Figure 2–59 presents a sectional motor with rotors connected through a flexible shaft.

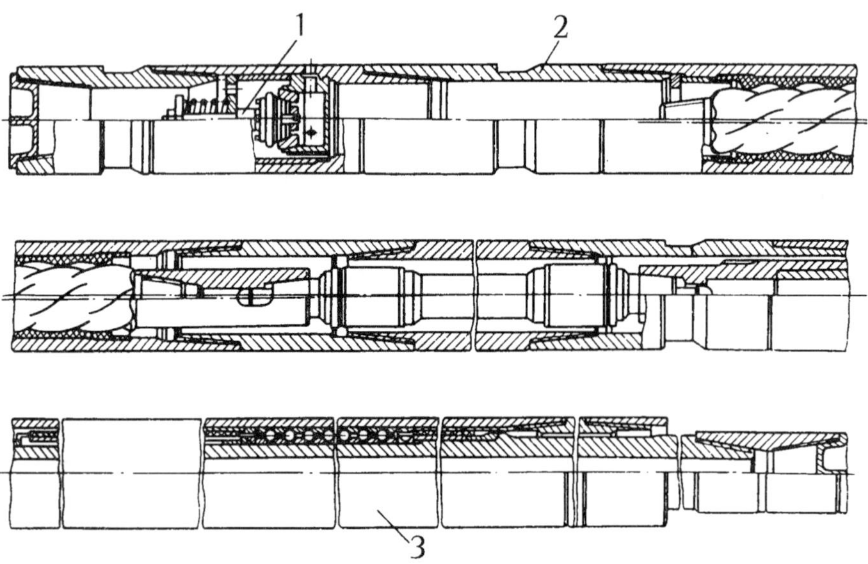

Fig. 2–57 The D1-195 motor design
1–relief valve, 2–motor section, 3–a spindle section

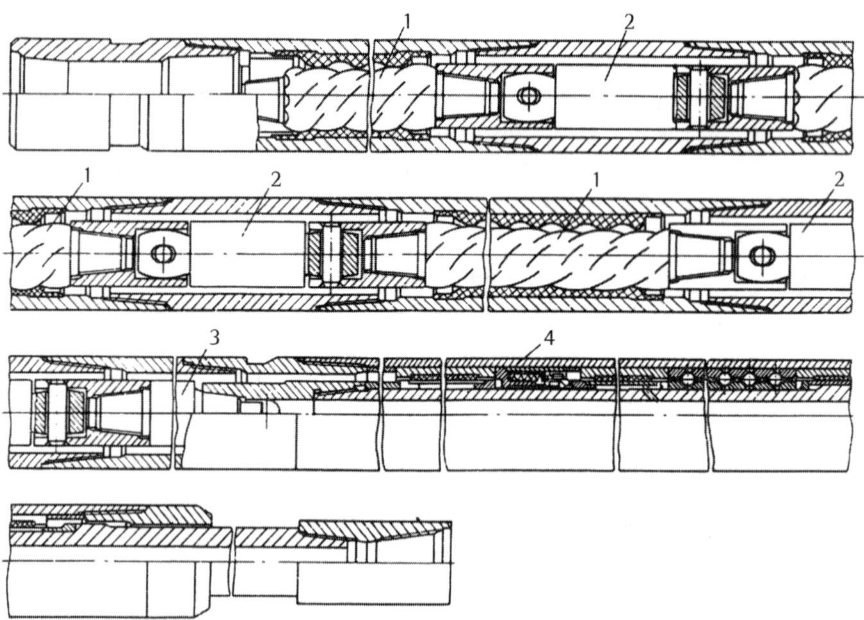

Fig. 2–58 The Sectional PDM with rotors connected through hinged joints
1–rotor, 2–hinge, 3–shaft, 4–spindle

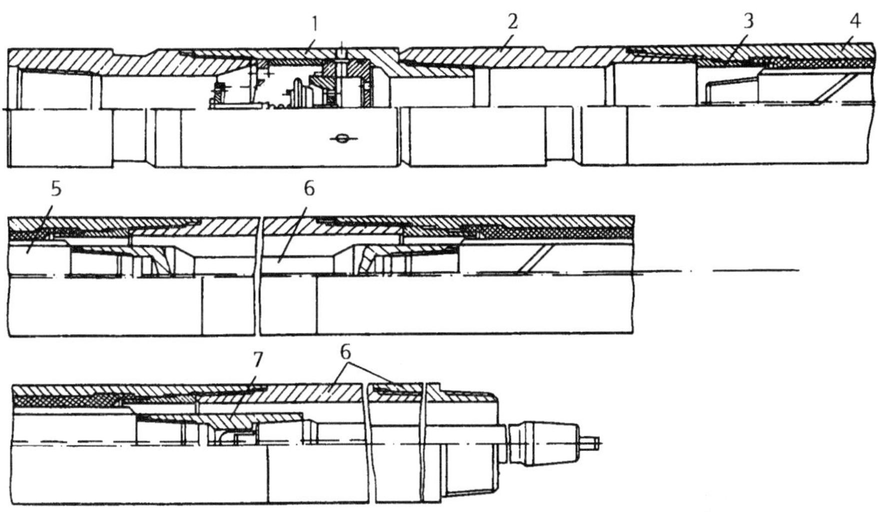

Fig. 2–59 Sectional motor with rotors connected through a flexible shaft
1–valve, 2–connecting sub, 3–spacer, 4–stator, 5–rotor, 6–crossovers, 7–clutch

Sectional PDMs performed as expected when drilling in complicated conditions with abrasive mud. In 1985, drillers were using 100 sectional motors, whereas in 1987 they were using 500 units. During this period, a total of 300,000 m was drilled in various regions of the country using the sectional motors.

The relief valve was another weak point in the equipment system used with a PDM. It was fitted above the motor and allowed the drilling mud to flow out of the DP into the annulus while tripping up. Figure 2–60 shows one of the valve designs. The valve tended to fail and required significant design improvement.

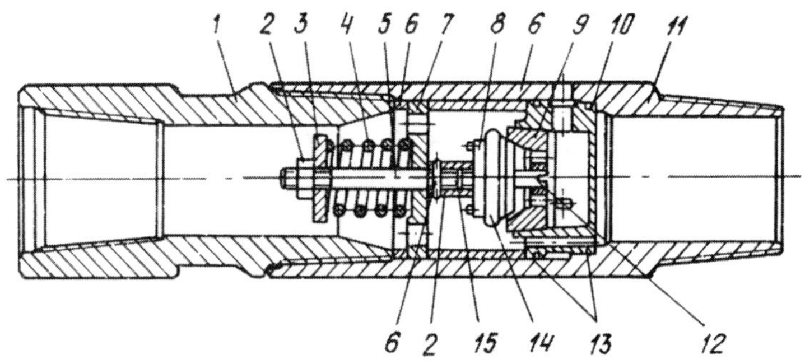

Fig. 2–60 Relief valve for D1-195

Workover and other PDM applications

PDMs found a broad sphere of application in well workover operations, such as drilling out cement or compacted sand plugs and removing salt and paraffin deposits from internal surfaces of casing and tubing strings. In the 1940, operators performed these jobs by running 127-mm turbodrills on a tubing string, [48] but these jobs could only be performed in a $6\frac{5}{8}$-in. or larger diameter casing string using special jars. This kept a low load level on the turbodrill.

Turbodrills such as the TS4A-104.5 and TS4A-127 that were built for these applications in the following years required a high mud circulation rate and pump pressure. Besides, they were high-speed and low-torque turbodrills. That is why oilfield personnel had to use rotary drilling methods in these operations. This consumed considerably more time and money because of the need to install additional equipment and use DP and was especially costly in offshore drilling. The engineers built PDMs with diameters of 54 mm, 85 mm, and 127 mm and fully resolved the problem for operations in both casing and tubing strings. The D-54, D-85, and D-127 DHMs are used nowadays for workover jobs in almost every producing region of Russia to provide a faster and cheaper workover method.

In summary, during the 40-year period after WWII, engineers managed to develop and build a series of HDHMs that met virtually every requirement of oil and gas drilling technologies (special types of these motors for new applications are described in the relevant chapters of the book. Yet, this does not mean that the work for their improvement has been finished. Emerging applications require continued development of new drilling technologies for this type of equipment.

Electrodrills [49] [50] [51]

Prehistory

The development of an electric downhole motor (EDM) is based on the idea of using electric energy as a more effective power for drilling operations. This kind of energy has the following advantages:

- It is cheap and convenient for long-distance transmission.

- It is easily transformed into other kinds of energy with high efficiency, allowing the use of automatic technologies and remote control.

- It maintains constant power by increasing the voltage to compensate for power losses in transmission lines.

All these reasons allow the development of highly effective DHMs for driving drillbits and automating the drilling process.

During the last 100 years, experts from the United States, Germany, France, Austria/Hungary, Romania, and Russia have tried to develop drilling machines with electric motors. The first efficient EDM for deep drilling was developed in the U.S.S.R. between 1937 and 1940. The desire to develop these motors was driven by the numerous advantages of the electric motor:

- power transmitted to the bit was high, constant, and independent of fluid flow conditions
- bit performance and rock destruction at the well bottom were controlled from the surface
- properties of the penetrated rock were conveyed by bit performance

Naturally, these advantages were accompanied by additional technical problems related to directing two parallel flows of energy to the bottom for bit rotation and bottomhole cleaning.

The first electrodrill consisted of a 3-phase, 4-pole motor with a capacity of 70 kW, a planetary gear that reduced rotation speed from 1450 to 363 rpm, and a spindle connecting the gear with a bit. The diameter and length of the electrodrill were 324 mm and 8.5 m respectively. The housing was filled with oil under pressure to protect the mechanism from the environment. The oil level was kept stable by a compensator. Drilling mud was pumped to the bit through an annular clearance between electrodrill case and the motor and mechanism housing.

Current was supplied through coaxial sections of a three-strand cable placed in DP. The cable sections were connected to each other by sleeves and pins when making up DP joints. At the same time an electro-differential bit feed regulator was developed.

In 1940 in the village of Kala near Baku, a well using this system was drilled for the first time anywhere in the world. The well was 1500 m deep. All parts of the EDM system passed this first field test. Based on this test result, the following conclusions were made:

- the electrodrill could operate successfully in an aggressive liquid medium under great hydrostatic pressure and dynamic load
- the system for current supply could use discrete power cables placed in DP
- substantial power at the bit improved bit performance, especially the penetration rate, compared to rotary drilling

Disadvantages uncovered during the tests were as follows:

- the electrodrill diameter was large, but the capacity was insufficient
- the planetary gear was a weak part of the system
- the cable connections broke down frequently
- the mud pump pressure increased due to the large diameter of the cable

Continued research was required to improve the electrodrill and the whole system. World War II suspended the research for six years. It was resumed in 1947 when the new non-gear 250 mm electrodrills were manufactured.

First 10 years of field applications experience

Electrodrills with cable section in pipe joints. Field tests performed in Azerbaijan during 1948 to 1950 and in Bashkiriya from 1950 to 1951 proved the effectiveness of the new electrodrill. In 1952, both test and commercial drilling started in Bashkiriya. By this time, a 215-mm non-gear electrodrill was developed. The capacity of the 215-mm electrodrill was 96–145 kW at 600–750 rpm correspondingly depending on the number of poles (10–8), whereas the capacity of the 250-mm electrodrill was 145–230 kW at 600–1000 rpm correspondingly depending on the number of poles (10–6).

Between 1952 and 1963, these electrodrills were used in the Ukraine, in the Kuibyshev (now Samara) area, and in Turkmeniya. By drilling about 500,000 m in field tests and commercial drilling for 10 years and in different geological conditions, the advantages and disadvantages of electrodrilling were fully evaluated. For a number of completed wells (mostly in Bashkiriya), the drilling performances were 15–20% higher compared to turbodrilling. However, these improved performances were not the rule and did not lead to the commercial application of electrodrilling. To apply electrodrilling widely, special repair shops with the required equipment and highly qualified electricians had to be in place. In addition, higher performance was limited by the following substantial disadvantages of electrodrilling.

The low resistance of cable connections and the frequent breakdowns of the current supply system. There were up to four breakdowns of a new current lead per 1200–1700 m of deep wells in Bashkiriya. During the depreciable life of a current lead (15 wells) there were 10 or more breakdowns. This resulted in idle round trips and substantial loss of time. The situation was much worse in deeper wells. A record of 319 current lead breakdowns was made while drilling well No. 157 (4368 m). The well was drilled in difficult geological conditions with weighted mud in the Azerbaijan Karadag-Damba oilfield. The time lost because of these breakdowns amounted to 5000 hours.

Substantial pressure losses, especially in connections of current lead sections in DP joints (the outer diameter of the cable connection and the tree-strand cable was 65 mm and 43 mm respectively). At a depth of more than 4000 m in well No. 157, pump pressure was 120–140 kg/cm^2 at a flow rate of 20–23 l/sec. In well No. 153, which was drilled by turbodrill in similar conditions, the pressure level was the same.

High rotation speed of the motor shaft dramatically decreased roller-cone bit life, especially in deep wells.

Electrodrill overhaul time was 25–30 hours in 1956, while in wells more than 4000 m deep it averaged 3–4 hours.

Small diameter electrodrills were not available for drilling lower intervals in deep wells and lateral branches in cased wells.

Cable in DP prevented the use of wireline instruments for BHA orientation. Efficiency was low for bent electrodrills when used for directional drilling. The same was true of the instruments used for its orientation (AOSU) as well as impulse inclinometers (type EE). For this reason, electrodrills were used only in vertical wells before 1963.

The technical and technological features of electrodrills were not taken into consideration. Competition with HDHMs and rotary drilling in shallow wells could not demonstrate the advantages of electrodrills.

These disadvantages were thoroughly studied and taken into account by experts and companies who were involved in the development and application of electrodrills. This paid off in future designs.

Cable electrodrilling without pipe. Between 1947 and 1954 in the U.S.S.R., a cable line system of electrodrilling was also developed. Bit reactive torque in this system was counterbalanced by the inertia of the electrodrill housing. The spindle shaft, together with the drilling bit, rotated alternately in both directions up to a preset rotation speed, while reversal was controlled from a surface board. A pump built into the electrodrill was used to wash the well bottom. The pump sucked mud from the well and discharged it through the bit to wash the bottom. Mud with cuttings flowed into a sludge pit, where it was cleaned and redirected into the well. Test drilling was performed in a shallow well in Krasnokamsk. However, this original and unusual system appeared inefficient because of extremely poor bit performance, which could not be compensated for by fast round trip operations to change the bit.

Improved commercial electrodrilling systems (1963–1970)

General characteristics of electrodrilling systems. Between 1963 and 1970, an electrodrill standard was worked out. It included specifications for electrodrills with diameters from 127 mm to 240 mm as well as optional equipment for commercial electrodrilling applications. The first commercial electrodrilling system is shown in Figure 2–61.

The system consists of the following basic units:

- electrodrill and other parts of the downhole DS assembly
- telemetric system (STE)
- DS with power cable
- automatic bit feed regulator
- control station and board
- transformer for the electrodrill power supply

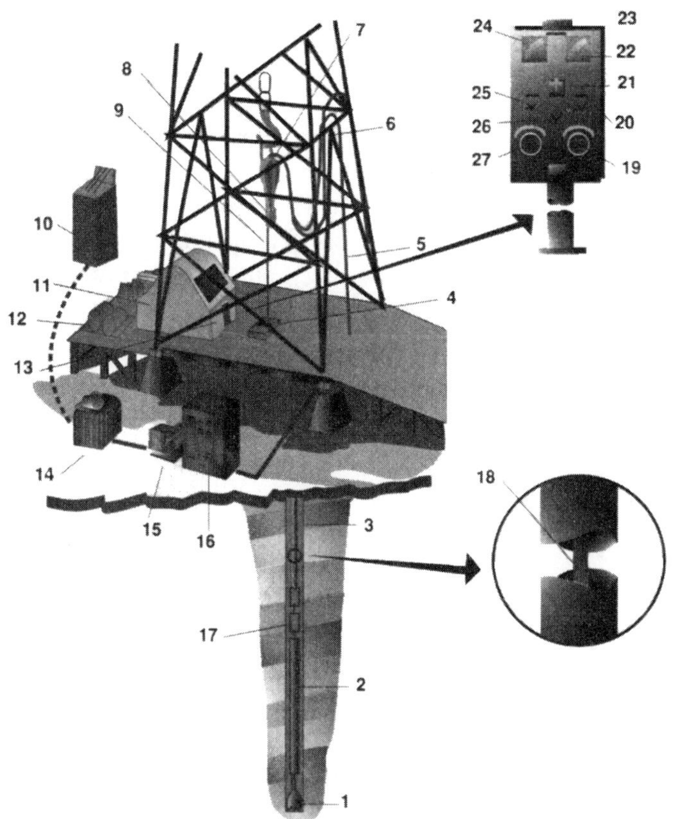

Fig. 2–61 Electrodrilling set-up

1– bit
2–electro drill
3–drillpipe string
4–rotory table
5–stationary external cable
6–flexible rubber insulated cable
7–swivel
8–collector unit
9–kelly
10–H.V. cubicle
11–drawworks
12–automatic electro-differential bit-feed regulator
13–electro-drill control board
14–power transformer
15–H.V. switch
16–electrodrill control station
17–tool joint
18–rubber-insulated feeder cable
19–bit-feed control unit
20–electrodrill motor control
21–selector switch for stepwise bit feed control
22–current meter
23–watt meter
24–volt meter
25–control for electro-differential bit feed regulator
26–slush pump control
27–electrodrill load-control unit

A conventional rotary drilling rig was used with a set of downhole and surface equipment prefabricated as attachments to the basic equipment. Surface equipment included special power transformers for electrodrilling, electrodrill control stations and boards, current collectors, an automatic bit feed regulator, gadgets for the electrodrill, and the current lead. In areas with no central power supply, a diesel generator unit was used.

Downhole equipment included electrodrills, electrodrill gear inserts, attachments for coring, check valves, devices for insulation inspection, mechanisms for well path control, tools for borehole path stabilization, logging tools, and telemetric systems (Fig. 2–62).

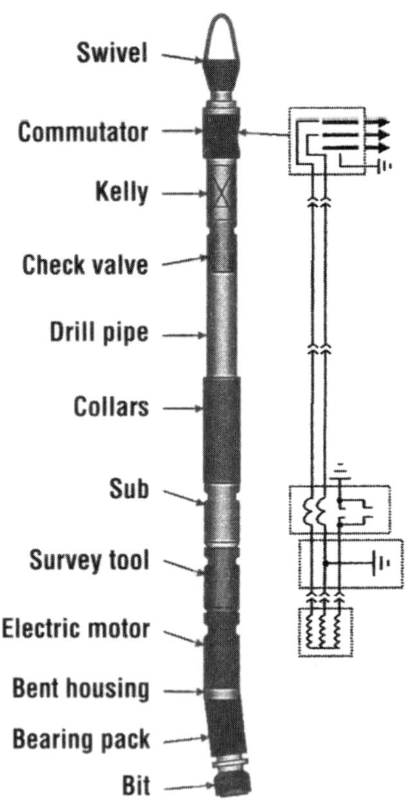

Fig. 2–62 Electrodrilling assembly

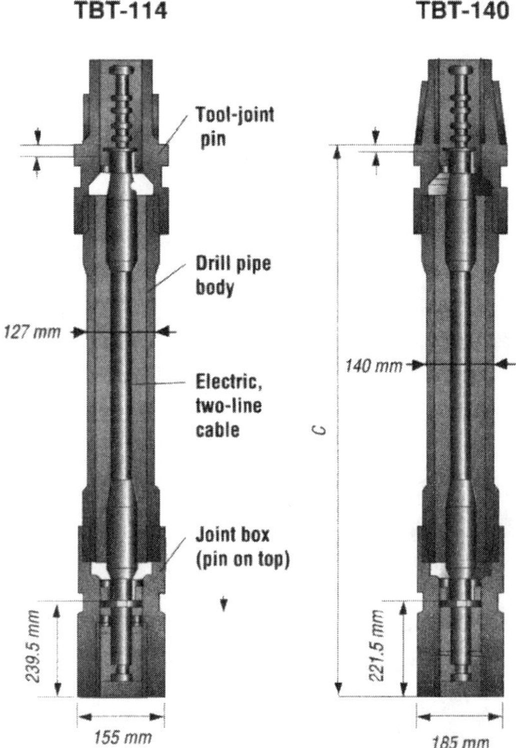

Fig. 2–63 Drillpipes with cable sections for electrodrilling

Telemetric systems were the most important part of an electrodrilling system. Together with an electrodrill, they provided a new advantage, i.e., drilling could be controlled based upon on-line information about the bottomhole drilling parameters.

Power was supplied to the electrodrill through a cable built into the DPs (Fig. 2–63). The cable led into a current collector (Fig. 2–64). The slipping contacts of the collector could rotate the DS if necessary.

Research carried out by the Kharkov SKTBE from 1963 to 1970 determined the design that is currently available for the electrodrill system. Proposals for further development of this system to improve efficiency, to develop DC applications, and to develop small diameter electrodrills did not find support in the U.S.S.R. or in other countries.

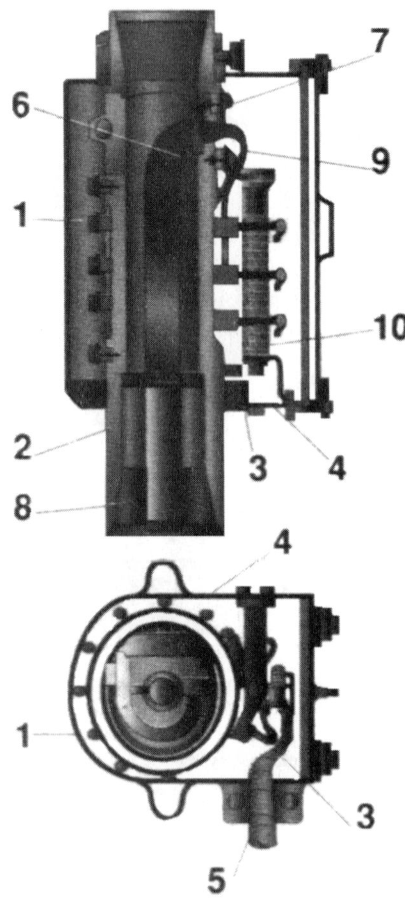

1–slip ring
2–collector body
3–stationary copper-graphite brushes
4–stationary collector housing
5–stationary external cable
6–feeder cable
7–flange
8–feeder line contact bushing
9–feeder cable strands
10–brush holder

Fig. 2–64 Current collector for electrodrilling

Electrodrills. An electrodrill (*see* Fig. 2–65 and Fig. 2–66) has a tubular body and consists of two basic units: a subsurface induction squirrel cage motor and an oil-filled spindle. If necessary, a mechanism for well path control and/or a gear insert (Fig. 2–67) for reducing bit rotation speed and for increasing electrodrill torque are installed between the motor and the spindle. Axial load from compressed part of the DS is transmitted through electrodrill housing and bypasses the motor rotor. Torque from the motor is transmitted to the bit through the spindle shaft. Specifications for the electrodrill are given in Table 2–38.

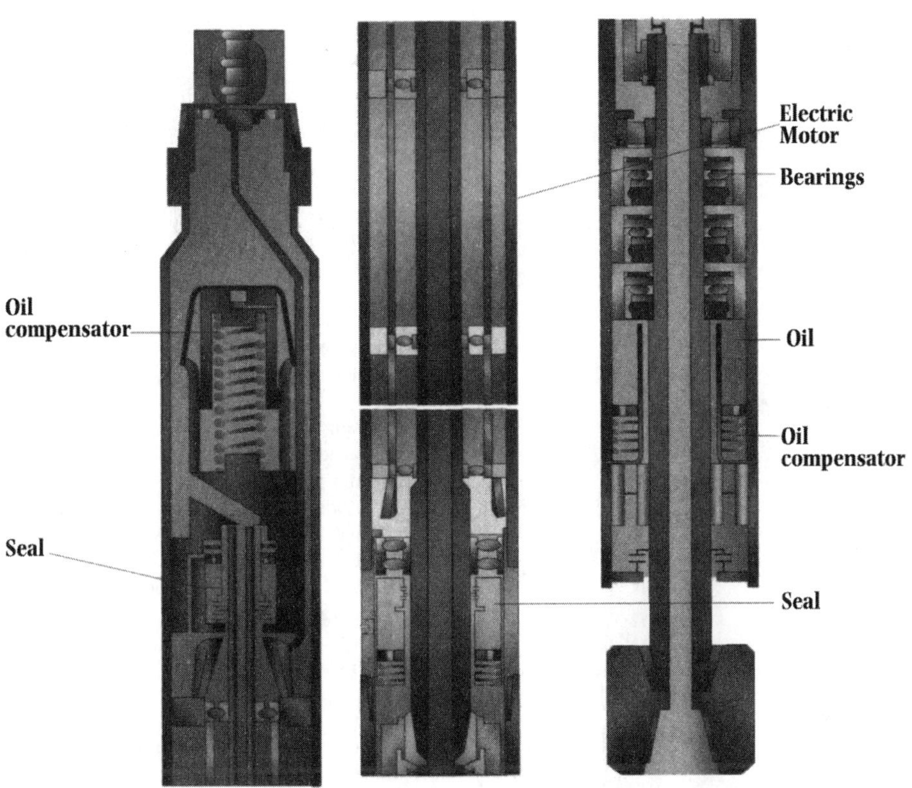

Fig. 2–65 Electrodrill schematics

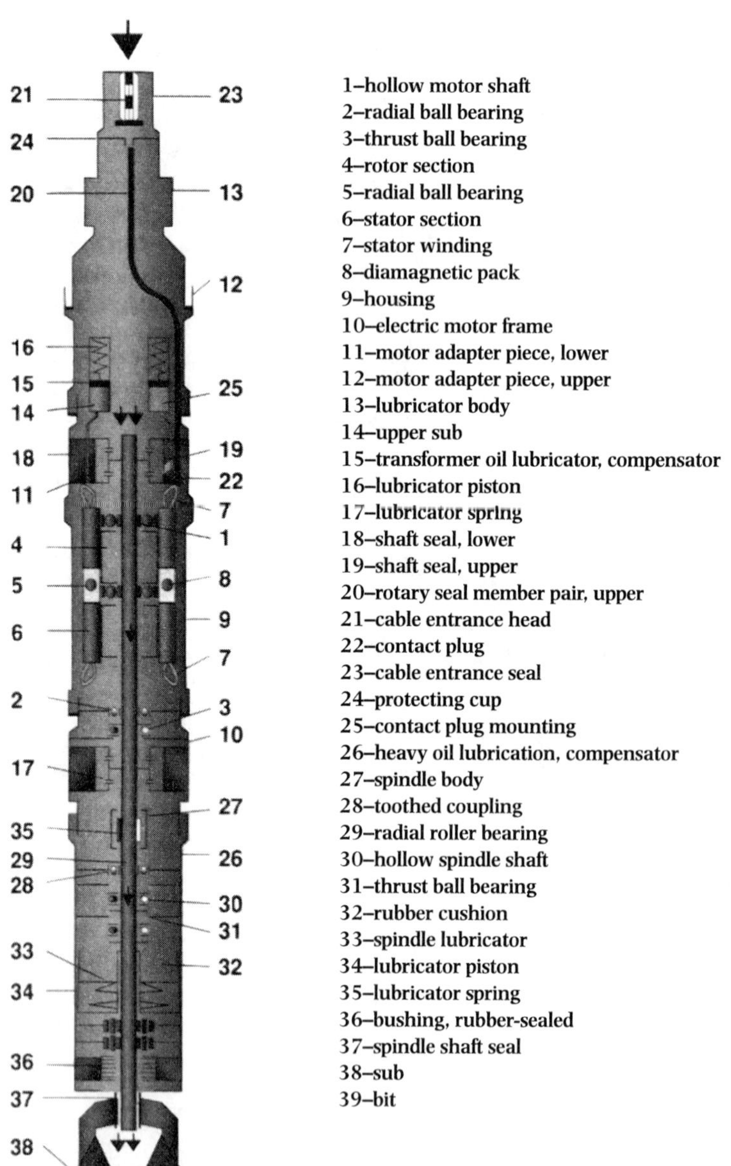

Fig. 2–66 Electrodrill design

TABLE 2–38
Commercial Electrodrills

Parameters: Electrodrill type	ER127-4B5	ER164-4M6B5	E164-8M7B5	ER190-4M4B5 (M1B5)	E190-8M7B5	E215-8M7B5	ER240-4M1B5	E240-8M7B5	ER290-6M7B5	E290-12AMB5
Power, kW	31,2	71	65	50 (110)	125	175	170	210	180	180
Voltage, V	750	1,350	1,100	1,000	1,300	1,550	1,350	1,700	1,550	1,750
Current, A	52	52	89	50 (107)	125	131	113	144	100	123
RPM*	430	140 /437	675	162 (162/477)	675	680	181 / 505	690	300	450
Torque*, Nm	330	4,100 / 1,500*	1,100	2,600 (5,300/2,160)	1,800	2,500	6,230 / 3,210	3,000	7,000	5,100
Efficiency	0.73	0.74	0.61	0.75	0.60	0.72	0.79	0.75	0.79	0.72
OD, mm	127	164	164	190	190	215	240	240	290	290
Length, mm	10,100	9,684	12,000	8,600 (11,450)	13,000	13,800	11,200	13,700	10,300	12,800
Mass, kg	850	1,100	1,500	1,400 (1,950)	2,200	2,920	3,000	3,640	3,700	4,600

* *Figures divided by slash correspond with different gear ratio (approximately 1:9 / 1:3).*

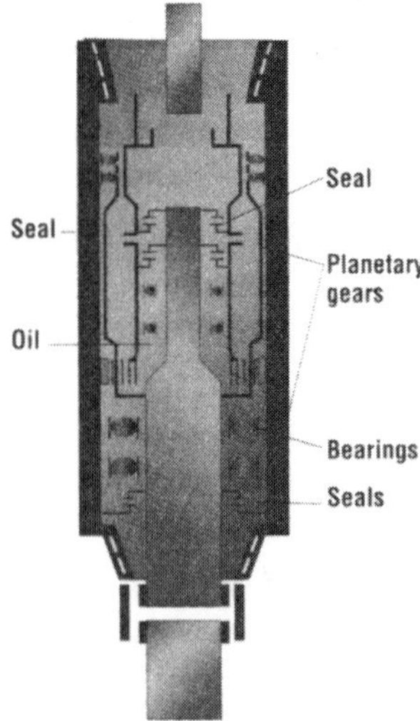

Fig. 2–67 Gear reduction box (insert) for electrodrill

Current lead. Power supply systems made up of two conductors in DP marked a substantial improvement in the current lead. Power to a three-phase electrodrill motor is supplied through a double-wire cable and DP (*see* Fig. 2–62). The diameter of the double-wire cable connection was reduced to 50 mm and had an oval cross-section of 35 x 15 mm. Pressure losses were reduced substantially when this cable was employed. Modernization of the cable connections sharply reduced the number of breakdowns in the current lead.

The current lead was designed for voltage up to 3000 V and for current up to 165 A with spikes up to 400 A for 3 seconds. Cable sections can operate at hydrostatic pressures of up to 115 MPa and at environmental temperatures of 100–130° C. Each section of the current lead cable ended with a contact bushing installed in a tool joint box on one side and with a current lead contact pin installed in a tool joint nipple on the other side. A safety sleeve protected the contact pin from damage (*see* Fig. 2–63).

Cable sections were used with 114-, 127- and 140-mm DP and with 129-mm ADP with internal upset ends.

Telemetric systems STE for directional and horizontal electrodrilling. A communication channel with the surface was the most important part of the STE to control downhole parameters. For the first time, telemetric systems for measuring geometric and geophysical parameters were applied in electrodrilling. This was made possible by using the DP as the communication channel for the cable. The (STE) was developed between 1966 and 1968. In 1968, it was used for directional electrodrilling in Bashkiriya for the first time worldwide and later in other areas of the U.S.S.R.

STE is designed to measure the geometric parameters of the well path geometric—the inclination, azimuth, and bent sub position. The STE consists of submersible and surface apparatus. The submersible part includes a survey meter that has a set of meter sensors and electronic apparatus. These sensors send electric signals that contain information on measured downhole parameters. The surface part consists of a receiving board for recording and visual inspection of subsurface parameters, and a connecting filter for connecting the receiving board to the electrodrill current lead and for separating the frequency of the power line current (50 Hz) from the system frequency. The survey meter is placed in a separate diamagnetic pipe, which is installed immediately above the electrodrill (*see* Fig. 2–62). The electrodrill current lead serves as a communication channel.

The telemetric system can measure inclinations up to 110°, azimuth and tool face from 0° to 360°. The margin of error for measurements is not more than 2.5%. Specifications for the STE are given in Table 2–39. The STE receiving board has an outlet for connecting standard instruments to record the measured parameters. It also has remote instruments installed near the driller's console. Design details and applications for the STE are presented in Chapter 4, Volume 2.

TABLE 2–39
Cable Telemetric Systems

System Type	Length, mm	OD, mm	Zenith Angle*	Azimuth and tool face*	Mass, kg
		For drilling with turbodrills and PDM			
STT - 108	6,300	108	0 - 60; 60 - 120	0 - 360	231
STT - 127	5,290	127			234
STT3P - 127M1G	8,195	172	-		695
STT3P - 190M1G	8,195	190			845
STT3P - 215M1G	8,195	215			942
		For electrodrilling			
1STE - 164U3	8,355	164			563
1STE - 185U3	8,355	185			967
1STE - 215U3	8,272	215			1065

* *Same parameters for all system types.*

Electrodrilling technology

General concept. In the early 1970s, the U.S.S.R. was the only country in the world where all three methods of drilling were used on a commercial scale, i.e., rotary, turbo- and electrodrilling. For this reason, it was important to estimate the ratio of application for each of the three drilling methods. As mentioned before, turbodrilling footage reached 80% of the total drilling volume in the U.S.S.R. by the late 1950s. As for electrodrills, discussion on expediency of their application has been going on until the present.

However, fundamental research of the key test (technological) wells (KTWs) carried out by the VNIIBT in the 1960s and the 1970s proved the suitability for all three drilling methods to be available to local drilling companies. The task was to apply each of the methods in those conditions where all the potential advantages could be used (*see* Chapter 3 for details). In that respect, electrodrilling in the FSU found its proper niche market. Examples of effective electrodrill applications are given as follows as well as in other sections of this book (*see* Chapter 3 in this volume and Chapters 4–6 in Volume 2.)

Deep drilling with weighted mud in Turkmeniya (1970–1980). The following wells were drilled with an optimal bit rotation speed and a mud density of 2.3 g/cm: (a) No. 808 with a depth of 5042 m and No. 809 with a depth of 4700 m in the Kotur Tepe oilfield; (b) No. 27 with a depth of 4803 m, No. 31 at 4616 m, and No. 32 at 5250 m in the Komsomolskaya oilfield. The last well was a record in terms of rates. Overall drilling rate was 1.8 times higher compared to well Nos. 52, 28 and 404 drilled by the rotary method.

According to data gathered from 1975 to 1976, the reliability index of the electrodrilling system increased substantially. Average mean-time-between-failures (MTBF) of electrodrills was 50 hours, the number of breakdowns of the current lead was 2.5 per 1000 m of boring while drilling deep wells with weighted mud.

Directional and branched lateral drilling in complicated geological conditions in Azerbaijan and the Ukraine (1960–1980). Well No. 1183 of about 4000 m in the Zagly-Zeiva oilfield, Azerbaijan, was drilled in steeply dipping formations where the angle of entry provided for a decrease in curvature. The well was drilled to the target due to continuous control of the azimuth and curvature.

Electrodrills were so successful in steeply dipping formations that in the western Ukraine more than 10 directional wells, which had been spud in by turbodrills, were drilled to the target by electrodrills. Exploratory multilateral well No. 801 was drilled at the Dolina oilfield, in the western Ukraine. In total, 12 horizontal and branched lateral wells including 11 development wells were drilled at Dolina. Initial daily production rates of the branched lateral wells reached 70–150 tons compared to 8–10 tons for vertical wells, because the latter wells could not always be directed to the project target in the given geological conditions.

Drilling with gaseous agents in complicated geological conditions (1970–1980). A classical example of using different circulating agents while electrodrilling is the directional 2813-m well No. 726 with a vertical deviation of 930 m in the Bitkov oilfield, the Western Ukraine. Surface hole drilling to a depth of 196 m was performed with foam due to lost circulation. An interval of hard rock from 196 to 1452 m was drilled with air, while mud circulation was used in the interval from 1452 m to 2813 m. The combination of electrodrills with the application of gaseous agents sped up drilling substantially and reduced the number of days in the drilling program by 52.

Interesting results were also obtained while electrodrilling a cluster of three directional wells (Nos. 1270, 1271 and 1272) in the Zagly-Zeiva oilfield, Azerbaijan, in 1977. The average depth of the wells was 2200 m. Two of the three wells were drilled with foam to a depth of 1100 m. Drilling rates of this cluster were much higher than of a similar cluster of wells (Nos. 1313, 1236 and 1237) drilled with turbodrills. The bit footage, penetration rate, and overall drilling rate were respectively 1.2, 1.7 and 1.67 times higher. It should be pointed out that the directional well path followed by the electrodrill with a STE is much more accurate than a turbodrill without a telemetric system.

Research and KTW drilling. Many years of experience in KTW drilling in the U.S.S.R. showed that electrodrilling was the most effective method for such operations (*see* Chapter 3). It provided the largest volume of information required for choosing optimal drilling methods and practices. Moreover, information obtained while KTW electrodrilling could be used successfully in other methods of drilling.

A good example of electrodrill research is a unique experiment drilling a deep well in a crystalline basement carried out in Bashkiriya. Borehole No. 2000 was drilled by electrodrill in 1964 in the southeast wing of the Tuimazi structure to investigate the earth crust in the Volga-Ural oil and gas province. The crystalline basement was drilled in the interval from 1798 to 4041 m. The basement was composed of inhomogeneous, frequently alternating rocks like gneiss, granite, and diorite.

In the interval between 2156 and 4036 m, researchers investigated the influence of drilling operating variables on bit performance. Drilling operating variables, capacity, current, and voltage consumed by the electrodrill were recorded in 100 research runs. Drilling was performed with E215/8 and E250/10 electrodrills using 295-mm TCI bits for hard rock. Motor shaft rotation speed was adjusted by the current frequency regulator. Drillbit rpm was varied from 285 to 680 by using electrodrills with different numbers of poles and by adjusting current frequency. Axial load varied from 50 to 300 KN. Researchers established the dependence of the bit penetration rate on axial load and rotation speed, as well as the optimal meaning of these parameters for maximum penetration per bit.

Horizontal drilling (1979–1990). When well No. 196 Uzibash was drilled in 1979, it was the first horizontal borehole drilled with an electrodrill in Bashkiriya. Conventional drilling equipment and tools were used. The kick-off point was between 1950 and 2130 m; the lateral section ran from 2130 to 2385 m; the maximum zenith angle was 102.4°; and the final vertical deviation was 607 m.

The drilling results from well No. 196 proved it was possible to drill horizontal wells with conventional electrodrilling equipment. However, the recommencement of horizontal electrodrilling did not occur until 1988. One of the wells was drilled in the New-Uzibash field and four experimental wells were drilled in the Lemesinskaya oilfield. Currently, electrodrilling is used in Bashkiriya and Tatariya where horizontal wells are drilled in pay formations with a thickness of 1.5–2 m. Another horizontal well, the Kotur-Tepe No. 1630, was drilled successfully under complicated geological conditions in Turkmeniya. This well reached a depth of 3653 m with a lateral section of 145 m. In the Salymskoye oilfield of Western Siberia, horizontal well No. 578 was drilled to a measured depth of 3330 m with more than 300 m of deviation. A 127-mm electrodrill was used in this well for the first time (*see* Volume 2, Chapter 4 for more details).

Future developments [52] [53]

Research in the 1980s showed that an optimal rotation speed range couldn't be provided in rotary drilling. The search for a new drive entered naturally into the conventional drilling technology. The PDM was adopted by those involved in drilling technology as the motor able to improve drilling practices and to control the well path in horizontal drilling. However, in spite of large investments and high quality research, the technique that was developed appears to be an imperfect solution. Hydraulic motors cannot provide the required range or smooth adjustment of rotation speed. Both the transmission of information and control of the well path in the hydraulic system are more expensive, less accurate, and have less communication capacity compared to a cable system.

In this situation, it was appropriate to consider developing an integral system based on electrodrilling to further advance progress in drilling. At first, development of electrodrilling was worthwhile in the construction of oil and gas wells in complicated geological conditions when gas, foam, and weighted mud with a density greater than 2000–2500 kg/m^3 were used. Even now, an electrodrilling system developed in the 1960s can successfully compete with any kind of modern drive when drilling with PDC bits.

A record of 12,000,000 m drilled proved the considerable potential of electrodrilling for deep, directional, horizontal, and multilateral drilling. The greatest advancement is expected in re-entry drilling including laterals from a cased parent, out-of-operation or marginal wells in old oilfields, and in offshore drilling.

Research and KTW drilling performed in different oil and gas fields of the U.S.S.R. from the 1960s through the 1980s were oriented to electrodrilling as a primary research method. KTW drilling promoted the development and adoption of the most effective drilling methods, practices, and machinery.

The field experience and research results gave sufficient information to encourage the use of electrodrilling as a well construction method. The extension of KTW drilling results allowed realistic proposals to use electrodrilling as a universal method for modern drilling. Electrodrilling combined the advantages of both rotary and DHM drilling and, at the same time, it lacked some of the disadvantages of the latter. Drilling experience in Bashkiriya, Turkmeniya, Azerbaijan, and in the Ukraine strongly suggested that electrodrilling provided savings of energy and materials. Electrodrill applications decreased environmental pollution and saved reservoir properties that in turn, reduced time to completion.

The DC electrodrill is very promising. This system uses the more reliable and space-saving single-conductor current lead, which decreases pressure losses in DP and permits the use of a fishing tool when necessary. In addition, the DC electrodrill can be adjusted smoothly across a wide range of rotation speeds. A DC electrodrill can be overloaded to a greater extent than an alternating current (AC) electrodrill, and this enables an increase in the net motor power.

Application of an integral cable instead of a discrete one is the other attractive aspect. This combines well with modern coiled tubing drilling (CTD) drilling technology. However, such a combination requires the development of a small (less than 5-in.) diameter electrodrill and research in the field of power cable implementation in CTD system. Recent studies, both in Russia and abroad, show positive prospects for the development of 3–4 inch diameter electrodrills for CTD applications.

Due to its characteristics, electrodrilling may find a wide application in the coal industry, mining, underground railways, laying trenchless pipe, and other branches of industry.

Electrodrill as a bottomhole transmitter

When considering electrodrilling prospects, it should be remembered that electrodrills were used in the framework of technologies developed for the rotary table and turbodrill as bit drives. These drives lack the main electrodrill advantage, meaning they do not optimize drilling operations under variable conditions.

An electrodrill serves a dual function (informative and executive), which is not accessible for either modern drilling systems or measurement while drilling (MWD). Signals created by special sensors (similar to MWD) are transmitted through the electrical communication channel. In addition, bit performance is evaluated by the signal strength.

An electrodrill serves as a sensitive transmitter of the bottom process and responds to all drilling alterations and deviations. It allows operational control of the drilling process. This advantage of electrodrilling creates the prerequisite to automate drilling.

A board of instruments was developed and manufactured by VNIIBT to record details of the electrodrilling technological and energy parameters. The board recorded voltage at the current collector, the current in each of the three phases, the active power consumed by the electrodrill, the reactive power, the axial load, the meterage (in time), the pressure in the manifold, and the drilling mud flow rate. In addition, watt-hour and var-hour m were installed in the board of instruments.

To control the drilling process, a telemetric system, STE1-I, was tested in the Shebelinskoye gas field in 1969. It generated data about the magnitude of the bottomhole axial load and load dynamics, as well as the amplitude and frequency of vibrations. Surface instruments recorded the dynamics of the bottomhole axial load (by oscillography), static (average) meaning of the bottomhole axial load, amplitude and frequency of the DS vibrations (oscillography using a sensor installed above the electrodrill), and other parameters.

Later telemetric systems STE-164, STE-185, and STE-215 were developed for directional drilling purposes and approved for batch manufacturing. This type of system records and shows the azimuth, zenith angle of the well, and bent sub position (*see* Chapter 4, Volume 2 for details).

The instrument metering board records the drilling technological and energy parameters and combines with the STE to acquire information to develop the correlation between the drilling indicators.

Bottomhole information obtained by STE can be used for different purposes, for example, to improve a drillbit run. The task is to keep the drilling operation variables that provide the maximum efficiency of by-bit energy consumption. The maximum penetration rate V_m at a similar power consumption N_p indicates the optimal mode of rock destruction. Most energy is spent for rock destruction, while the remaining smaller amount of energy is wasted on drillbit destruction and wear as well as other unproductive work.

Electrodrilling controls an effective rock bit break-in procedure and determines the end of the bit run. This is made possible by on-line recording of phase current magnitude I_1, I_2, I_3 and power N. Variations in the records help develop standard procedures for effective drillbit runs in specific geological cross-sections. The combination of the loading mode and optimal bit rotation speed that were determined by a frequency unit considerably increased drillbit performance.

It is possible to judge a drilling process by the character of power and current records. A smooth record indicates an optimal mode of rock destruction. Vibrations of the DS produce small peaks. When a bit bearing is worn out, power and current grow dramatically and the records are characterized by sharp peaks. On occasion, the current protection switches off the electrodrill. That is how it is possible to determine bit condition by the character of records, the end of bit run, and bit bearing failure.

An electrodrill determines a more precise meaning of WOB. As it is known, WOB measurement errors can be significant, especially in highly deviated and horizontal boreholes. As the magnitude of power N_p solely depends on the magnitude of WOB, it is possible to determine the WOB accurately by adjusting the load through the meaning of N_p: $N_p = b\ x\ (WOB)$, where $WOB = N_p/b$. The coefficient of proportionality b is in direct relationship with the specific torque of rock destruction M_s. M_s can be determined in experiments with WOB by adjusting $M_y = 975(N_{p1}\text{-}N_{p2})/(WOB_1\text{-}WOB_2)n$, where n is rpm.

Adjusting the load by 2–5 times, it is possible to determine the magnitude of specific torque M_s and, therefore, the full magnitude of WOB during one bit run.

The magnitude of the specific torque M_s in the well depth function determines adequately drilled rock. If the composition of drilled rock is known, better control of bottomhole destruction is available from the electrodrill's on-line information.

Adaptation of this technology helps to develop new advantages in modern electrodrilling. The advantages are connected with seeking new solutions based on the theory of bottomhole boring and destruction. Practical MWD possibilities are largely based on a comparison of the drilled section with experience from wells drilled earlier. This empirical approach significantly limits the possibilities of further scientific and technical progress in drilling. That is why drilling practice inevitably has begun to change in favor of a more wide application of electrodrilling, which was ahead of its time. Now this system should be perfected, and new electrodrilling applications should be put into practice.

Conclusions

Electrodrilling technology can be considered a commercial drilling method similar to the worldwide use of rotary and HDHMs technology. Electrodrilling combines some of the advantages of both rotary and HDHM methods, including:

- a large range of drillbit rotation speed
- independence of energy transmitted to the drillbit from the fluid flow
- use of different borehole cleaning agents
- controlled drilling of deviated and horizontal boreholes

An electrodrilling system is effectively applied in complicated geological conditions and old reservoirs where weighted mud or different mud mixtures must be used.

The application of electrodrills is very promising in directional, horizontal, multilateral, and research boreholes.

EDM for CTD is one of the best opportunities to overcome the current CTD problems. That is the area where all the disadvantages of EDM with standard single joint pipe vanish. On the other hand, the best cable electrodrilling ideas could be realized in CTD, so we could now be back on the right path.

References

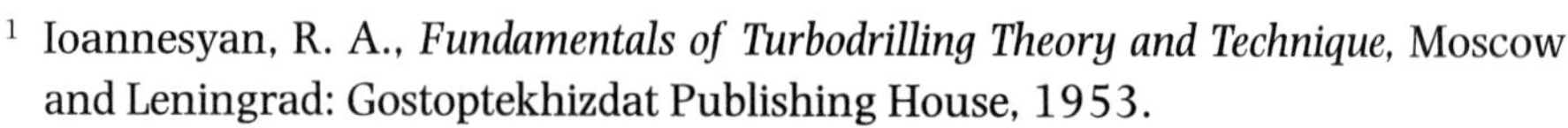

[1] Ioannesyan, R. A., *Fundamentals of Turbodrilling Theory and Technique*, Moscow and Leningrad: Gostoptekhizdat Publishing House, 1953.

[2] Ibid.

[3] Ibid.

[4] Antonov, N. V., Ya. A. Gelfgat, and M. T. Gusman, "Chapter XXIV, Oil Well Drilling Guide," *Turbodrilling*, Moscow and Leningrad: Gostoptekhizdat Publishing House, 1947.

[5] Shumilov, P. P., *Oil Well Turbodrilling*, Volume I and II, U.S.S.R.: ONTI NKTP, 1936.

[6] Shumilov, P. P., *Fundamentals of Turbodrilling Theory*, Gostoptekhizdat, 1943.

[7] Shumilov, P. P., Oil Well Turbodrilling: Selected Works, Moscow: Nedra Publishing House, 1968.

[8] Ibid.

[9] Ioannesyan, 1953.

[10] Shumilov, 1968.

[11] Ibid.

[12] Antonov, 1947.

[13] Shumilov, 1943.

[14] Gusman, M. T., B. G. Lyubimov, G. M. Nikitin, I. V. Sobkina, and V. P. Shumilov. *Calculation, Design, and Operation of Turbodrills*. Moscow: Nedra Publishing House, 1976.

[15] Ibid.

[16] Ibid.

[17] Antonov, 1947.

[18] Shumilov, 1968.

[19] Shumilov, 1968.

[20] Shumilov, 1968.

[21] Gelfgat, Ya. A., A. V. Orlov, G. M. Finkilshtein (VNIIBT), A. S. Shafutin, and M. N. Yadulayev (AzNIPI), "Summary Results of Drilling Test Wells in the Karadag-Damba Field," *VNIIBT Transactions, Issue XIV.* Moscow: Nedra Publishing House, 1965.

[22] Gelfgat, Ya. A., "Turbodrilling Application Experience in Heavy Mud Conditions," *Neftyanoye Khozyaistvo (Oil Industry) Magazine,* No. 8, Moscow, 1953.

[23] Gusman, 1976.

[24] Kurepin, V. I., F. N. Fomenko, and G. S. Gevorkov, "Study of Operating Regimes of Electrodrills Used in Combination with Diamond Bits on Prikarpatburneft Company oil-rigs," *Neftyanoye Khozyaistvo (Oil Industry) Magazine,* No. 9, Moscow, 1969.

[25] Barshai, G. S., and N. I. Buyanovsky, *Theory and Practice of Turbodrilling,* Moscow: Gostoptekhizdat Publishing House, 1961.

[26] Litvyak, V. A., , L. I. Brai, and V. F. Ryzhenko, "The Results of Turbodrills with Floating Stators Testing," *Neftyanoye Khozyaistvo (Oil Industry) Magazine,* No. 8, Moscow, 1984.

[27] Fomenko, F. N., *Boreholes Drilling With Electrodrills,* Moscow: Nedra Publishing House 1974.

[28] Vadetskiy, Yu. V., N. D. Nikomarov, and N. D. Derkach, "The Significance and Prospects of Low-speed Down-Hole-Motor in Well Drilling Technological

Progress," *Neftyanoye Khozyaistvo (Oil Industry) Magazine,* No. 1, Moscow: Nedra, 1975.

[29] Sabirzyanov, A. K., V. M. Safarov, and N. G. Anikin, "Reduction Gear Turbodrill TR2Sh-195 Commercial Tests in Zapsibburneft Oil Company," *Bureniye (Drilling) Magazine,* No. 10, Moscow, VNIIOENG, 1972.

[30] Derkach, N. D. and E. N. Krutik, "Industry Designs of Reduction Gear Turbodrills," *Oil and Gas Wells Construction Onshore and Offshore, Proceedings VNIIOENG,* Issue No. 2–3, Moscow, 1992.

[31] Derkach, N. D. and E. N. Krutik, "Gear Reduction Turbodrills Improve Drilling Results" *SPE 49258,* SPE Annual Technical Conference and Exhibition, New Orleans, Louisiana, USA, 27–30 September, 1998.

[32] Ibid.

[33] Ibid.

[34] European Commission, *New Solutions in Energy Supply–Heat Resistant Gear Reduction Turbodrills* (OG/201/98/DE/UK/RU), Energy publication series No. 234, Brussels, 2000.

[35] Budyanskyi, V. S. and S. Yu. Brudnyi-Chelyadinov, "Modular Turbine-Screw Motors," *Neftyanoye Khozyaistvo (Oil Industry) Magazine,* No.1, Moscow: Nedra, 1993.

[36] Lyubimov, B. G., A. N. Shindin, and V. P. Shumilov, "On the Problem of Turbine Design with reduced Axial Dimensions for Hydraulic Down-hole Motors," *Oil and Gas Wells Construction Onshore and Offshore, Proceedings VNIIOENG,* Issue No. 2–3, Moscow, p.30–35, 1992.

[37] Gusman, M. T., D. F. Baldenko, A. M. Kochnev, and S. S. Nikomarov, *Downhole Screw Motors for Boreholes Drilling,* Moscow: Nedra Publishing 1981.

[38] Gusman, M. T. and D. F. Baldenko, "Screw Downhole Motors," *Bureniye (Drilling),* Issue #6, Moscow: VNIIOENG, 1972.

[39] Baldenko, D. F., F. D. Baldenko, and A. N. Gnoevykh, *Screw Downhole Motors,* Moscow: Nedra Publishing, 1999.

[40] Gusman, 1972.

[41] Baldenko, 1999.

[42] Ibid.

[43] Gusman, 1981.

[44] Vadetskiy, Yu. V., M. T. Gusman, D. F. Baldenko, and S. S. Nikomarov, "The Prospects of Downhole Screw Motors Application in 11th Five-year Plan," *Neftyanoye Khozyaistvo (Oil Industry) Magazine*, No.11, Moscow: Nedra, 1981.

[45] Ibid.

[46] Gusman 1981.

[47] Baldenko 1999.

[48] Antonov, 1947.

[49] Abysbayev, B .I., N. K. Baibakov, Y. A. Gelfgat, and M .Y. Gelfgat, "Electrodrilling: Past Experience and Present Opportunities" *SPE 38624*, SPE Annual Conference, San Antonio, USA, October 6–8, 1997.

[50] Abysbayev, B. I., N. K. Baibakov, Y. A. Gelfgat, and M. Y. Gelfgat, "Electro drill provides alternative drilling system, *Oil & Gas Journal*, Feb. 9 1998.

[51] Ibid.

[52] Abysbayev, B. I. and B. V. Baidyuk, "Study on the Drillbit WOB-RPM Rational Correspondence and the Tasks on Intensive Drilling Technology Development," *VNIIBT Proceedings*, Issue 66. Moscow: Nedra, 1988.

[53] Abysbayev, B. I. and B. V. Baidyuk, "Electrodrill as a Transmitter of Downhole Information and a Research Tool," *Proceedings of Russian Scientific Conference on the Oil and Gas Basic Problems*, vol. 3, Moscow, 1996.

3

WELL DRILLING OPTIMIZATION METHODS IN THE FSU

Mission Statement and Substantiation of the Necessity for Developing a New Method of Well Drilling Technology Optimization

In the late 1950s and during the 1960s, the oil industry in the FSU witnessed a significant growth in the number and depth of oil and gas wells being drilled. For example, the total drilled footage of oil and gas wells in 1961 was 8,360,000 m. By 1971 the footage increased by 42% (11,800,000 m), and by 1974 it had increased 71% (300,000 m). In addition, the average development well depth increased from 1792 m in 1961 to 1980 m in 1971 and 2012 m in 1974. Exploratory wells grew from 1995 m to 2554 m and 2675 m for these same years. During this 15-year period, the overall drilling rate increased 20% from 544 m/rig-month to 652 m/rig-month, whereas cost per meter went from 87.7 rubles to 162.0 rubles, an 85% increase.

In 1961, 18 deep wells (greater than 4500 m) were drilled, whereas 125 deep wells were drilled in 1973. During the same period, the annual penetration volume of deep wells increased from 85,000 m to 611,000 m with an average overall drilling rate of 200–300 m/rig-month and an average depth of 4750 to 4890 m.

This data indicates that the growth of oil and gas production was impeded by the slow growth rate of overall drilling speeds and resulted in a significant increase in cost per foot.

Lack of a comprehensive approach to optimizing well drilling technology was one of the main reasons for the low drilling rate. Meanwhile, such issues as the timely and valid selection of drillbits, optimum parameters of drilling methods and practices, selection of a DHM, well design, and required equipment and tools became more important. At that time, solutions for these issues were determined from experience by analyzing data from hundreds of drilled wells. This approach resulted in the development of numerous facts about the oilfields and oil provinces before the drilling techniques were in place. For many years, such techniques were only applied in very large fields.

The existing methods of designing drilling techniques (a detailed description follows) included a number of drawbacks and were based mainly on the analysis of statistical data from bit runs in offset wells and adjoining fields. Unlike the statistical data accumulated by Western drilling contractors, the information obtained by Russian companies was not as objective. As a result, progress in optimizing drilling technology in the FSU was rather slow and ineffective.

Our investigation and analysis of the available data concluded that the process of optimized well drilling in the United States was also based on statistical information from previously drilled wells or from individual bit runs, and it allowed for prompt adjustment of drilling parameters. We believe the efficiency of this approach was due to a number of features that relate to organizational and technical issues, including:

- a consistently high level of drilling technology and equipment provided by equipment and material supply companies
- availability and smooth operation of instrumentation and recording equipment for drilling operations
- exclusive use of the rotary drilling method that facilitates the optimization of drilling parameters within the acceptable range of rotational bit speed variation
- a reliable system of payment for drilling crews that was not dependent on drilling results

These circumstances contributed extensively to the objectivity of the statistical information from the rigs that served as a basis for drilling process optimization. In the FSU, the statistical data was based on drilling reports by foremen and was not reliable. It could not be used as the only source of information or the basis for optimizing the drilling process because:

- Russian rigs did not have sufficiently advanced and reliable working equipment to achieve or maintain high quality drilling operations.
- Russian rigs lacked reliable instrumentation, recording equipment, as well as maintenance tools or systems.
- Russian drillers used several drilling methods, and the most widely used method, turbodrilling, did not produce information about rotational bit speed.
- The method of payment for drilling crews significantly affected information in the official reports and produced a lack of objectivity; hence it could not be used as the only source of data to plan drilling process optimization.

These circumstances led us to the conclusion that in order to obtain objective and valid information for drilling parameter optimization, we would have to drill specially designated wells. This work was started by the VNIIBT in 1961.

The plan provided for drilling certain wells in oilfields under the control and guidance of the VNIIBT specialists who would be free from the negative factors typical of a wide-scale drilling operation. Next, it employed an integrated approach to developing well design and actual well drilling. The plan provided for efficient use of modern drilling equipment and techniques. Specific conditions of the optimized drilling program were elaborated and implemented during the wide-scale development of a field. As mentioned in Chapter 1, this was made possible because, unlike the West, the FSU could employ a single contractor to conduct development drilling on individual fields throughout the entire life of the field. In a very few cases, two or more drilling companies participated in the development of large-size unique fields. Even in these cases, the work of each drilling contractor was limited to assigned sectors of the field.

The principal difference between the drilling program for these types of wells and the numerous "fast" or "demonstration" wells was that scientific and research work was built into the drilling program. One of the goals of the plan was to collect comprehensive information for use in further development, improvement, and implementation of drilling technologies.

Therefore, several years after a large number of experimental wells had been drilled with the participation of the VNIIBT specialists, the wells that were used for accumulation of drilling experience and relevant information were KTW, in contrast to key stratigraphic wells that were aimed at obtaining geological information. Later on, the name *KTW* became generally accepted and received official status as well as the right to be used in publications. It was also registered in the Mining Encyclopedia as a specific type of oil and gas well. [1]

KTW Drilling Principles and Procedures for Implementation

In accordance with the latest drilling procedure dated 1981, the name *KTW* was given to a well on which an active test was performed throughout the entire well borehole length, or at certain individual intervals, with the goal of collecting basic primary data required to adjust existing wells or develop new well designs and drilling procedures.

KTW were normally drilled on promising large fields where delineation drilling was carried out to obtain enough information in the early stages of field development to support selection of the following items using the most advanced equipment available for the drilling industry:

- well design
- bits and DHM types and designs
- drilling method
- optimum drilling practices
- BHA type
- methods of preventing and eliminating downhole problems
- other integral elements of the drilling process

On partially drilled oilfields that were in operation but had a sufficient volume of drilling remaining, KTW were constructed using an additional drilling program with new drilling equipment and technologies.

Drilling companies always carried out KTW construction in cooperation with NIPI, which developed feasibility studies for KTW drilling in a specific field and also provided scientific and technical support and guidance to drilling operations. To realize that support, the measurement and control instruments and recording equipment were installed on the rigs. The term *rig* is used here and throughout the book to mean a complex of derrick, drilling equipment, BOP stack, mud circulation and cleaning system, power supply, and other facilities necessary for well construction. NIPI conducted the necessary tests and studies and also processed and analyzed test results.

Specialists recommended using rotary and electrodrilling methods to drill KTW because these methods ensured compliance with the main conditions of tests such as independence of the main drilling parameters (WOB and rpm) from each other and from the drilling fluid circulation rate and properties. They also provided the required range of bit rpm; however, this recommendation did not preclude use of hydraulic DHMs for KTW drilling.

Optimized drilling parameters that were obtained when drilling the KTW with electrodrills were used in later development drilling of fields that employed electrodrills as well as turbodrills and PDMs. In these cases, the motor type and characteristics could be specified along the intervals in geological sections of the well, which maintained the recommended drilling parameters. Intervals where best results were obtained using rotary drilling were drilled using this method and the same drilling parameters.

The KTW drilling operations were financed in accordance with the "Method of Additional Cost Calculation for Construction of the KTW." This document was issued by the Ministry of Oil Industry on September 10, 1975, and approved by the Department of Cost Estimate Norms and Price Setting in Construction Work within Gosstroi (State Committee on Construction) of the FSU. In 1968, specialists from the VNIIBT institute developed and published the first version of the *KTW Drilling Procedure*.[2] In 1971, after incorporating comments and suggestions from various organizations, a second version of the *Procedure* was published. [3] The first two editions were aimed primarily at optimizing borehole deepening technology, i.e., selection of a drilling method and parameters based on information obtained about bit type, BHA, and DHM for various well intervals.

However, experience from the growing volume of KTW drilling data in various regions indicated that the KTW drilling program should not be confined merely to optimizing the well borehole deepening process. On their own initiative, research institutes and drilling companies in the regions carried out KTW drilling to identify the most efficient methods of eliminating drilling problems, selecting well design, and other elements of the drilling process. As a result of this work, a third enlarged edition of the *KTW Drilling Procedures* was published in 1976. The third edition included a number of specific techniques for obtaining the required information on the entire well drilling process such as:

- selecting well design
- developing classified well geological and technological sections
- selecting bit types
- determining optimum types of drilling mud and drilling parameters
- eliminating drilling problems
- drilling in a productive horizon
- other elements of the well drilling process[4]

Specific techniques developed by G. M. Finkelshtein for optimizing drilling practices with blade (drag) bits were described for the first time in an attachment to this edition.

Methods of KTW Drilling was reworked and published in 1982 as a shortened guideline (document designation RD-39-2-642-81). The document was approved by the Ministry of Oil Industry and was in force through 1987.

Development of a Mathematical Model of Well Deepening and Its Use in KTW Drilling

Analysis of existing optimization techniques of the well-deepening process

Based on a study of more than 100 publications, techniques (the term *techniques* being defined as design techniques, optimization techniques, or methods) for drilling parameters in Russia and in the West were divided into two groups:

1. Techniques developed on the basis of laboratory and bench tests including destruction studies of rock specimens
2. Techniques based on full-scale tests and studies of well drilling processes

   The second group included:

   a. techniques for effective and prompt determination of optimum drilling parameters and drillbit pull-out moment for a specific bit run, drilling method, bit type, and other design conditions

   b. techniques based on statistical data from previously drilled wells that was analyzed and processed using various methods (this analysis has been done using computers during the past 20 to 25 years)

   c. techniques based on drilling process optimization using data obtained from active tests

The drilling process optimization techniques of "c" type were developed using the empiric basic dependencies between the bit performance results and controlled parameters of a drilling process. Further study of these dependencies with the purpose to find extremum, allowed the determination of optimum values of controlled drilling parameters that took into account all factors that affect the well deepening rate for specific geological conditions in a given well.

The first group includes the technique that was the most widely used in the FSU and was developed by a group of scientists led by professors L. A. Shreiner, N. N. Pavlova, and B. V. Baidyuk. The technique was based on rock hardness analysis using indentation of a flat bottom cylindrical die. Development of this technique made it possible to perform prompt analysis and evaluation of the entire complex of rock properties that affect the drilling process. [5] [6]

These studies paved the way for investigating physical and mechanical rock properties in the oilfield sections. For example, they first introduced the concept of rock surface and volumetric (or solid) failures dependent on the applied axial static load. The rock properties, which were determined by the studies, were generalized and used to forecast drillbit performance. However, because these dependencies were separated from the real drilling process, they were used for overall classification of geological sections as applied to the fundamental selection of drilling technology and equipment. The principal importance of this technique was that it provided an opportunity for a large-scale study and the practical application of physical and mechanical rock properties in drilling operations for various regions of the country.

Another technique used in the first group was one developed by professors R. M. Eigeles and R. V. Strekalova. [7] It was used to calculate the penetration rate by utilizing results of bench tests to determine the dependence of rock failure strength on the penetration depth of an individual bit tooth when taking dynamic loading into account. However, the system of equations designated to determine drillbit operating features only reflected the entire bit operation schematically. Furthermore, it did not reflect the bit interaction with drill mud that was important for the rock destruction process at the bottomhole, the pattern of change over time, the performance of the bit cutting structure and bearings, etc.

A group of techniques that were based on field data found wider utilization because they had the advantage of taking into account all factors that affected the drilling process under field conditions, and these could not be imitated using lab and bench tests.

The techniques for prompt determination of optimum drilling parameters for specific bit runs, were, in turn, subdivided into two main groups, the "Model" and "Prospecting." [8]

The first group of techniques took into account the necessity of step changing bit weight P at constant bit rpm n to establish the dependence of penetration rate V_m from P, and at constant P establish the dependence of V_m from n by changing the latter parameter. While determining the effect of these variables on bit durability, the optimum values of controlled drilling parameters were identified to achieve the best level of optimization criteria. The parameters selected for the latter were mainly maximum penetration per bit and minimum cost per foot. The maximum bit run speed was seldom used. Various researchers used different numbers of test runs. The combinations and matrix tables built up the number of methods.

Yet because the tests and studies were done while drilling standard commercial wells, their results were limited and did not always allow optimizing well borehole deepening through the entire length of the borehole.

The type of prompt optimization previously described was used in the FSU by such researchers as V. S. Fedorov, [9] M. P. Gulizade, [10] G. D. Brevdo, [11] and others. In the United States, the best known techniques in that area were developed by scientists and researchers E. M. Galle, H. B. Woods, [12] [13] [14] F. S. Young,[15] [16] and M. Bingham. [17]

Due to their large number, not all of the "prompt modeling" techniques could be mentioned in this book, but they were applied and used by drilling companies in the FSU and drilling contractors in the United States. However, in the FSU they were used primarily in exploratory drilling, which is explained by some of the following negative aspects:

1. These techniques were developed mainly for rotary drilling method, which dominated in exploratory, especially deep wells, and had a limited rpm range.
2. As mentioned previously, several parameters were selected for optimization criteria such as maximum penetration per bit run and minimum cost per foot. Maximum bit run speed was seldom used as an alternative criterion, which, as shown later in this chapter, is more acceptable for large-scale development of well drilling in a field.
3. A significant amount of additional cost and time for the research work was unacceptable during commercial drilling, which did not allow finding the extremum for these dependencies to establish optimum levels of the main controlled drilling parameters.
4. The most serious drawback was the fact that the tests and studies were performed using equipment and technology designed for commercial drilling such as drillbit selections, drilling fluids, circulating rate, and other factors that affect bit performance results. This complex of parameters must be specifically selected to achieve optimization of the well borehole deepening process.

The drawbacks, noted in items 3 and 4, were also true for the second group of techniques for prompt determination of the optimum drilling parameters, the "prospecting" group. Rather than constructing mathematical models of a well-deepening process, these techniques were based on test drilling during one bit run and registering the levels of penetration rate and drilling time with this bit that were considered optimum at each individual combination of bit weight and rpm. Parameters such as minimum cost per foot for a particular bit run, maximum bit running speed, and maximum penetration rate may be used for this criterion.

A large number of scientists and researchers in both the FSU and United States participated in development, testing, and utilization of prospecting methods of the well borehole deepening optimization. Among them were M. A. Fingerit, E. A. Volgemut, [18] M. G. Eskin, [19] E. A. Kozlovsky, [20] Gulizade, [21] and other scientists in the FSU as well as American scientists such as C. D. Rodgers, [22] V. Edelberg, [23] Bingham, A. Lubinsky, and others.

It is difficult to make comparisons between the efficiency levels of the two groups of techniques for prompt determination of optimum drilling parameters. However, it is worth mentioning that these techniques were quite useful, especially for prospective and exploratory drilling, when there was little information about the geological sections of wells that were being drilled.

At the same time, during routine development drilling in an explored field, the efficiency of both groups of on-the-fly techniques was relatively low compared to the techniques that were based on the entire complex of factors used for drilling special key wells. From our point of view, that method may provide no more than 10 to 15% of the total effect of complex optimization processes. Still, they could be useful in exploratory drilling.

While describing the techniques based on the use of statistical data from previously drilled wells, it should be said that they once played a positive role in the arrangement of the data documenting system for results of bit runs and helped to identify the best ones that provided a basis for development of new well designs. In the last 10 to 15 years, such data has been processed and analyzed using modern computer equipment and software.

In the FSU, these optimization techniques were developed and used in the 1950s and 1960s by a number of scientists and research engineers. Among them were N. I. Shatsov from the Moscow Institute of Oil and Gas (MING), now Moscow Oil and Gas Academy, Fedorov from the Petroleum Institute in Grozny, and, in later years, professors Eigeles and Strekalova. A. S. Bronzov, V. I. Volfson and A. M. Yasashin from VNIIBT dealt with techniques within the Automated System for Control of a Drilling Process (ASUT) development.

One of the essential drawbacks to these techniques was the fact that they were based on processing the results of the "passive" experiments, which required a large volume of statistical data and covered a relatively small range of independent variable factors.

In addition, the original documentation in the FSU that was used for selecting drilling parameters was insufficient in the majority of cases. Another reason is that a large amount of drilling used turbines, and the operating data did not include parameters such as bit rpm.

As previously mentioned, drilling process optimization in the United States is based on information from previously drilled wells. The methods of drilling operations used in the United States obtained high quality original data; however, recent publications indicate that these methods also had certain disadvantages. For example, the first part of an article titled "Drilling optimization: If it ain't broke, fix it!" in *World Oil* [24] magazine contained the following introduction, "The frustrations and limitations of trial and error planning, even with experience from offset wells, are formidable."

The authors of this article cited data published earlier in paper number SPE 15362 presented by J. F. Brett and K. K. Millheim at a 1986 conference in New Orleans and related to the "Learning Curve" theory. The theory was illustrated by information from the study of 2010 deep wells drilled in various fields in four different regions around the world. The study indicated that as soon as the number of drilled wells increased from 1 to 11, drilling time decreased almost twofold.

Probably the issue of learning was important for American companies because exploration and development drilling in a field was done by several drilling contractors rather than by one drilling company. Some of them were spud-in new wells in fields where other contractors had already drilled wells. Moreover, they had to get new experience and information from drilling at these fields.

Nevertheless, their experience also proved that improvement and optimization of drilling technology used in a certain fields took quite a long time. One of the key statements from the article was quite specific: "Of all the factors that retard an organization's movement toward optimized drilling, the worst is inability to iterate enough cases."

The second part of this article titled "Drilling optimization: Practice makes perfect" was an attempt to present a computer's ability to improve the optimization process. The idea was based on a software application that could help find solutions. The system contained modules for hydraulic analysis, drag and torque, casing design, etc. The well planning process became an iterative computer simulation of situations used to find optimal decisions. So, the conclusion again was that "Optimized drilling is approached through trial, implementation, and evaluation experience gained from drilling similar wells of those drilled in a given area."

With that background, we would like to consider the most substantiated and promising technique of well borehole deepening, the one that uses data obtained from various active experiments to determine the basic dependence of bit performance results from controlled drilling parameters throughout the entire length of the borehole. This technique was developed by taking into account all factors affecting the well borehole deepening process and included further study of these dependencies' extremum to determine the optimum levels of controlled drilling parameters. However, such field research work can only be done using special experiments and KTWs drilled after or during outpost wells drilled in hydrocarbon fields, which includes required information about the geological section.

The program for drilling these wells should provide for additional financing of the work and should not have a restricted time frame. In the 1960s and the 1970s, specialists from VNIIBT implemented a method of well-deepening process optimization. The program planned to recoup additional investments later, which proved to be the case.

Test studies and development of a mathematical model of the well-deepening process

Three regions were selected for drilling the first test wells using the three methods (rotary, turbodrilling, and electrodrilling). All these regions featured different geological conditions and well depths and had the infrastructure and specialists required to do the work. Particular attention was paid to the availability of the service bases for repairing electrodrills and the personnel experienced with electrodrilling, since, as mentioned earlier, this type of DHM fit the planned test work best.

The three regions selected for the test well drilling were the Samara region, Azerbaijan, and Bashkiriya. In Azerbaijan, the program provided for drilling wells with a design depth of 5000 m in the Karadag and Karadag-Damba field (the field had a manmade dam that encompassed part of the field located beneath the Caspian Sea). In terms of geology, the field was similar to many adjacent fields, which, later on allowed a comparison to the test work results. Two fields, the Dmitrovsky and the Sosnovsky, were chosen for the test well drilling program in the Samara region, and in Bashkiriya, exploratory test wells were drilled in the Duvaiskaya Zona field as well as the exploration company activity area.

The researchers who performed the test program were to identify the following four functional dependencies in order to build the well borehole deepening model:

1) $V_m = f_1(P)$ at (n) and Q - const;
2) $V_m = f_2(n)$ at P and Q - const;
3) $T = F_1(P)$ at P and Q - const;
4) $T = F_2(n)$ at (n) and Q - const

where

V_m is the drilling penetration rate
T is the bit on bottom time
P is the WOB
n is the bit rpm
Q is the drilling mud circulation rate

These are conditional on a complete and thorough bottomhole cleaning by selecting the required Q and the corresponding velocity of drilling fluid coming out of the bit nozzle and the nozzle's cleaning pattern.

Two methods were used to determine the first of the dependencies:

1. A drillbit was pre-selected for the test drilling based on information about the type of rock in a certain well interval. The bit was broken-in for about four to five minutes at low bit weight. Next, bit weight was continuously increased according to a pre-designed pattern in steps of equal time intervals, or after drilling every 0.5 m, it was increased up to the level that provided maximum V_m. The penetrated section length and time were measured for these short stepped intervals.

These operations were carried out in the beginning, middle, and end of a bit run. When the penetrated well intervals were composed of homogenous rocks, the desired dependence was built in rectangular coordinates using the information received without any additional data. If the interval was composed of intercalating rocks with various physical and mechanical properties, establishing the dependence required the use of geophysical log data and identifying stringers with equal durability levels. [25]

2. The method was suggested by the American research scientist Lubinsky. The method provides for the use of Hooke's law for an elastic rod type deformation of a DS because of changing bit weight:

$$\Delta \ell = \frac{\Delta P \ell}{EF}$$

where

$\Delta\ell$ is the longitudinal strain of a compressed DS from a drillbit operation, m

ΔP is the bit weight change, kg

ℓ is the DS length, m

F is the DP cross-section area, cm^2

E is the modulus of elasticity, kg/cm^2

Elongation ($\Delta\ell$) of the compressed string during a period of time (Δt) is an average drilling penetration rate during the same period (Δt):

$$V_m = \frac{\Delta\ell}{\Delta t} = \frac{\Delta P}{\Delta t} \bullet \frac{\ell}{EF} = V_t \bullet \frac{\ell}{EF} = CVt$$

where

V_t is the bit weight change speed that determines the penetration rate, which is directly proportional because of the effects of Hooke's law

The following procedure was used for the test:

- bit weight was increased to the required level
- drawworks drum was put on brake

- time of bit unloading at each point was measured using a weight indicator

The data from these measurements was used to build bit weight variation in the time curve and calculate the penetration rate, which allowed building a diagram of its dependence from the bit weight.

One of the advantages of this method is the ability to determine the dependence $V_m=f_1(P)$ in a short interval (0.1 to 0.3 m), which is important when drilling wells in areas with frequent intercalated stringers of different drillability.

The method was used in the Samara region. A total of 25 measurements were taken in various horizons of the Upper Carboniferous and Bavlinian suites at a bit weight variation of 5 to 14 tons and rotary speeds of 75 to 90 rpm while drilling with a 161-mm bit. Figure 3–1 presents one of the diagrams built with distinctly delimitated sections and corresponding to various rock failure modes: surface (1), fatigue (2), and volumetric (3).

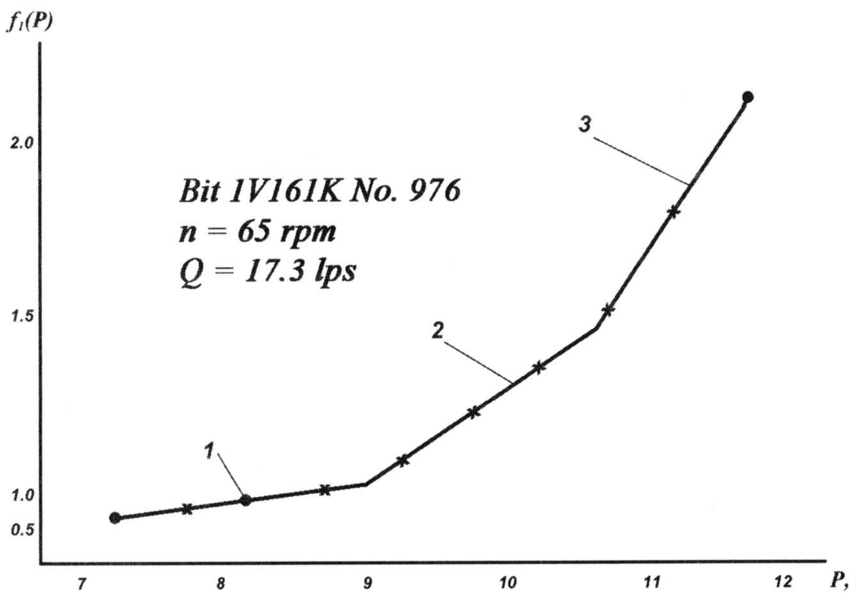

Fig. 3–1: Dependence (P) determined upon A. Lubinsky's method (1) surface, (2) fatigue, and (3) volumetric rock fracture modes

The second dependence $V_m=f_2(n)$ for the rotary drilling method was determined using the same technique that was applied to find dependence $V_m=f_1(P)$, i.e., a step change of bit rotational speed at constant P.

The third dependence of bit durability from bit weight, all other parameters being constant, was determined using the results from a number of bit runs at various bit rotational speeds and weights, with each weight level constant during drilling. For this purpose, intervals with the same drillable level were selected whenever possible.

Dependence of bit durability from rotational speed was also determined using results from a number of bit runs at various rotational speeds and bit weights that were constant during one run.

While determining the third and fourth dependencies, the researchers simultaneously studied the penetration rate variation at the time of each bit run with other drilling parameters remaining constant.

The drillers also used electrodrills such as ES215/2 (two-section), E215/10, and E215/8 for drilling test wells, which allowed them to control the shaft rotational speed. At 50 Hz, the E215/8 and E215/10 electrodrill rotational speed levels were 680 rpm and 530 rpm respectively, whereas with a frequency converter, the rotational speed levels were 380 rpm and 450 rpm at the corresponding frequency levels of 28 Hz and 34.5 Hz. A two-section electrodrill at 450 rpm was used to drill the test well in combination with a frequency converter at 34.5 Hz. The researchers determined all dependencies and functions using the same method that was applied when drilling the rotary test well except in cases where Lubinsky's method was used.

For the latter method, the DS must be rotated to eliminate the effect of a DS hanging up because of friction against the borehole walls. However, at that time, constant DS rotation while running electrodrills was not desirable because it negatively affected the condition of the power and data transmitting cables.

The study using turbodrills was more difficult. Nevertheless, it was conducted using turbine tachometers, regulating the drilling fluid circulation rate, and maintaining the rate within acceptable limits to have complete bottomhole cleaning. However, main dependencies used for building the model were based on the data from drilling test wells using rotary and electrodrilling methods.

In Azerbaijan, test well No. 198 in the Karadag and Karadag-Damba fields was drilled with an electrodrill. In addition, three KTWs, Nos. 156, 157, and 153, were drilled using the rotary, electrodrilling, and turbodrilling methods respectively. Researchers also used information from commercial well No. 166 that was drilled using a rotary method.

Test well drilling in the Dmitriyevsky field of the Samara region and an adjacent field that featured a similar geological section included KTW Nos. 156, 90, and 157 using rotary drill method, well Nos. 154 and 90 with turbodrills, and No. 168 with an electrodrill. KTW No. 403 in the Sosnovsky field was drilled using rotary drill method and small diameter bits.

In the Tuimazinsky field in Bashkiriya, test well Nos. 1283, 1249, and 1524 were drilled using electrodrilling. Well No. 39 in the Duvansky Zone field was drilled using rotary and turbodrilling methods with 161-mm drillbits.

The research drilling in all these regions commenced in 1961 and was completed in 1964 and 1965 with a total of 20 wells drilled in various geological conditions at a wide range of well depths. This test drilling obtained the principal basic dependencies listed previously.

For optimization criteria, the researchers took the maximum penetration per bit run and the maximum bit run speed. It is worth mentioning the reasons that the minimum cost per foot was not selected as an optimization criterion. We believe this criterion is only acceptable for evaluating the efficiency of drilling one well or a small group of wells because the drilling penetration rate usually decreased after drilling costs were minimized.

For example, this occurred while drilling wells with depths of 2000 to 2500 m using rotary and turbodrilling methods in the Tatariya and Bashkiriya regions of the Urals-and-Volga oil province. Normally, wells drilled using the rotary method were cheaper than similar wells drilled with DHMs, yet in the latter case the well deepening was faster. From the cost efficiency point of view, however, the method that allowed a faster penetration rate was more advantageous when drilling a large number of wells in a field or region. This is because it enabled faster field development while using the same number of drilling rigs, otherwise the number of rigs would have to be increased to achieve the same field development rate. In either case, total drilling cost is lower when using the method that provides faster penetration rate. This concept was based on specific calculations and was published in an article in "*Neftyanoye Khozyaistvo*" (*Oil Economy and Management*) *Magazine*. [26]

When the organizational and technical support conditions of a drilling process were equal, penetration rate was the factor that affected overall drilling speed. Therefore, it was selected as a criterion for KTW drilling optimization since such a well would serve as a model for large-scale development drilling in the field.

To determine optimum levels of the main drilling parameters, such as bit weight and rotational speed, equations of basic functions $V_m=f_1(P,n)$ and $T=f_2(P,n)$ were developed using the results of drilling experimental and KTWs and were analyzed for their extremum. Assumed results of bit runs were calculated using the optimum parameter levels that were determined and that allowed achieving maximum values of the selected criteria. Next, validity of these assumed bit results was checked by drilling development wells. This method of building a mathematical model of a well-deepening process was used as a basis for the corresponding section of all editions of the *Techniques of KTW Drilling*. The effect of factors such as dynamic load, differential pressure, bottomhole cleaning pattern, etc. had not been studied yet. They were reflected in the basic functions as constants and experimentally determined from drilling KTW.

Basic dependencies identified using the results of the test well drilling in the three regions mentioned previously, have the following form.

Average penetration rate dependence from bit weight and rotational speed:

$$V_{max} = k \cdot n^{\alpha}(P - P_o)^m \qquad 3.1$$

where

- n is bit rotational speed, rpm
- P is the bit weight, tons
- P_o is $V_m = f(P)$ straight line X-intercept, characterizing the moment of a rock failure pattern change during a transition from the surface-and-fatigue (sections 1 and 2 in Figure 3–1) to the volumetric failure (section 3 in Figure 3–1)
- k is the proportionality factor, depending on a level of bottomhole cleaning and bit tooth wear rate
- m is the exponent, depending on physical and mechanical characteristics of drilled rock, bit type, and level of bottomhole cleaning; in most cases it is ≈ or = 1

Test well drilling in the Karadag field in Baku [27] indicated that in similar geological conditions, the exponent m equaled 1 for the majority of bit runs. Rotational speeds varied from 70 rpm in well No. 156, drilled using the rotary method, to 300, 380, 450, and 680 rpm in well No. 198, drilled with an electrodrill. In addition, while drilling well No. 156, the researchers clearly determined that this regularity occurred only at bit weight levels below 14 to 15 tons when standard bits with low drilling fluid velocity at jet outlets were used. Use of jet nozzle bits ensured a direct proportion between V_m and P at bit weight levels up to 26 tons, which obviously proved the dependence of the exponent m on the quality of bottomhole cleaning.

Figure 3–2 shows the dependence $V_m=f(P)$ determined while drilling well No. 198 using an electrodrill with a rotational speed of 680 rpm. Similar dependencies were found at rotational speed levels of 300, 380, 450, and 530 rpm. All tests were carried out in the interval 2684–3832 m in the Sabunchinsky and Balakhinsky suites of Cenozoic productive thickness. The rock that made up this interval could be characterized as having medium hardness. The lithological composition was quite homogenous.

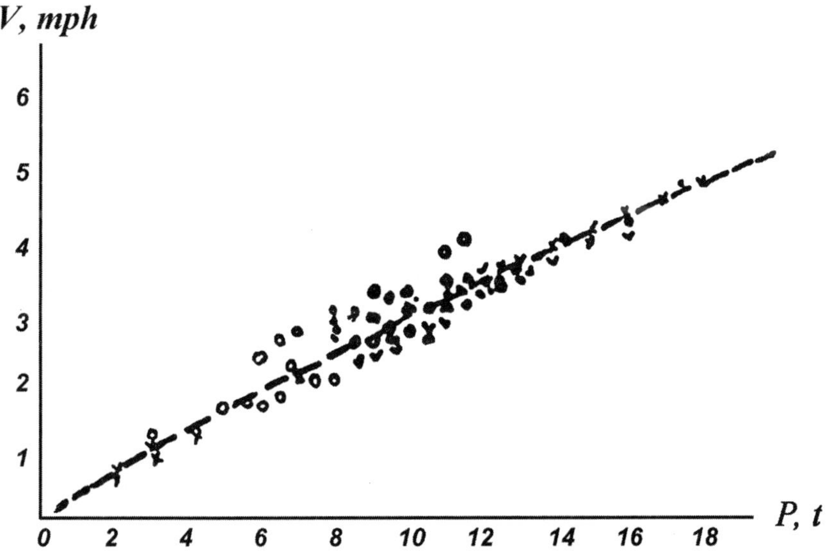

Fig. 3–2 Penetration rate dependence from WOB, Karadag field (Azerbaijan), well No. 198, electrodrill E 215/8; type B11S bit rotational speed of 680 rpm; interval 2684–2952m in the Sabunchinsky suite (the measurements within one bit run are indicated with the same symbol)

Information obtained from these tests and studies agreed with the conclusions of Galle and Woods who used field test results to show that m=0.6 for very soft rock and m=1 for the remaining rock categories. This conclusion was also supported by the results of studies made while drilling KTW Nos. 90 and No. 157 in the Samara region using the rotary method and well No. 168 using an electrodrill (Fig. 3–3 and 3–4). [28] These wells were drilled in the Dmitrovsky field that featured a cross-section composed of medium, hard, and very hard rock in Paleozoic deposits from Upper Carboniferous to Turonian, Frasnian, and Kynovian horizons.

The α exponent depends on the bit type, physical and mechanical properties of the rock, quality of bottomhole cleaning, and differential pressure level Δp at the bottomhole. In the majority of cases, its level is below 1. Sometimes it equals 1 at $\Delta p \rightarrow 0$, for example, in air drilling.

The tests performed to determine the dependence $V_m = f_2(n)$ indicated it did not conform with the linear law, and the curve reached a plateau when n increased (Fig. 3–5). [29]

The actual level of α was determined by taking the logarithm of Equation 3.2, determined in two points at actual levels of V_m, at similar and different loads. [30]

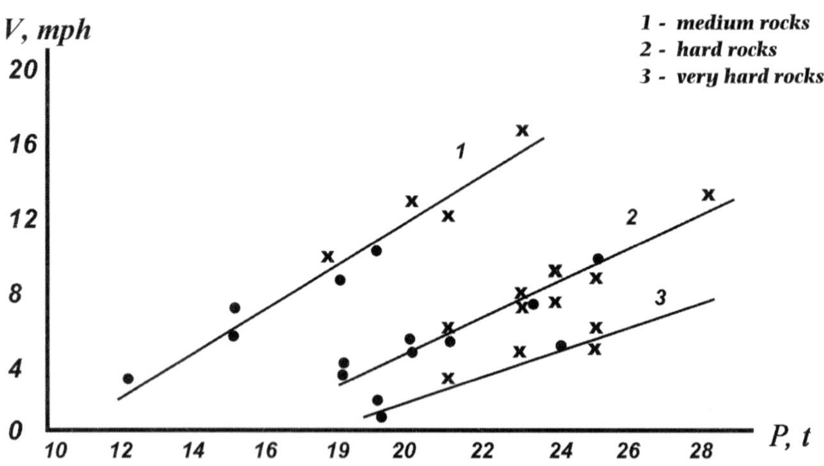

Fig. 3–3 Penetration rate dependence from WOB while drilling rocks of different strengths at bit rotational speed of n=60 rpm–Dmitrovsky field, Samara region, wells No. 90 and No. 157, Interval 557-2,494 m, up to 2000–2200 m drilling using water circulation

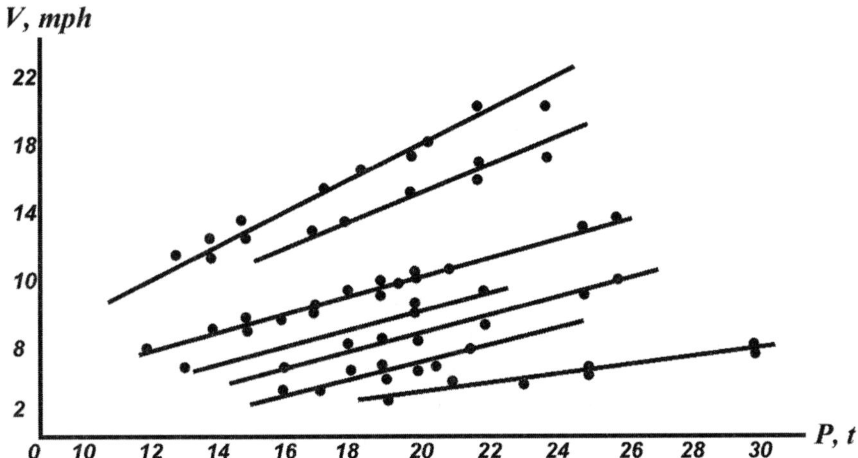

Fig. 3–4 Penetration rate dependence from WOB while drilling rocks of different strengths (electrodrill E 215/8, bit rotational speed of 680 rpm)

The researchers found that the main reasons for the decrease of α were the increase of rock hardness and well depth, especially Δp (differential pressure). These were the primary reasons rather than a concept popular at the time that attributed a decrease in deepening the level per bit revolution to the lack of sufficient time for a bit tooth to make contact with the rock. This did occur, but it was not a primary reason for the decrease. In the Karadag field, α varied from 0.2–0.3 to 0.9–1.0.

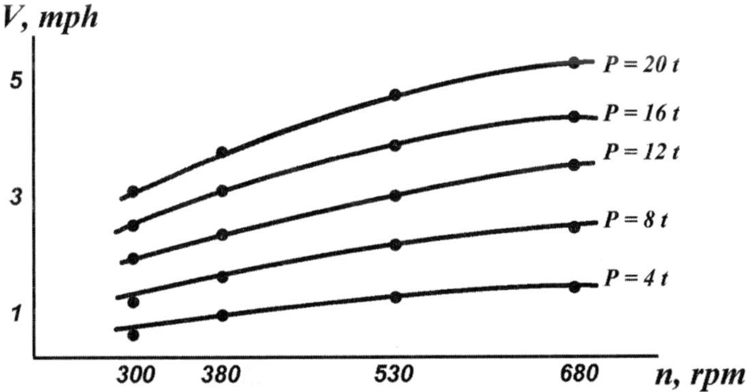

Fig. 3–5 Penetration rate dependence from bit rotational speed while drilling at different levels of WOB, Karadag field (Azerbaijan), well No. 198, electrodrills E 215/8, E 215/10, ES 215/2 in combination with frequency converter; Sabunchinsky and Balakhinsky suites; interval 2684–3832 m

Since the level of P was significantly greater than P_0 in most cases and P_0 was close to zero when drilling rocks that were easy to drill, Equation 3.1 was used in the following form:

$$V_{max}=K{\bullet}n^{\alpha}P^{m} \quad 3.2$$

Equation 3.1 should be still used for very hard rock, when P_0 is quite high.

To summarize the information about determining V_m dependence from P, the apparent contradictions between Figures 3–1, 3–2, 3–3, and 3–4 should be explained. It is worth mentioning that in accordance with the modern views expressed by Bingham in Figures 3–1 and 3–3, the range of the $V_m=f(P)$ dependence plateau has not been reached due to insufficient bottomhole cleaning, whereas the pattern of rock failure at the bottomhole in Figure 3–4 dominated because of the high bit rotational speed. For the same reason, volumetric and volumetric-fatigue rock failure in Figure 3–4 probably was not observed. In a quite generalized approach used by Galle and Woods and also in our studies, exponent m and the presence or absence of P_0 were among the features accounted for in this case.

Bit time on bottom dependence from bit weight and rotational speed (bit bearings failure):

$$T= T_0 - ap - бn \quad 3.3$$

where

- T is bit on bottom time in hours (bearing durability)
- T_0 is the formal empirical parameter characterizing original durability of tri-cone bit bearings, depending on bit size and operating conditions; its variation is directly proportional to a bit diameter
- a, б are coefficients, determined for specific drilling conditions, characterizing the intensity of T variation depending on $P{\bullet}$ and n

Equation 3.3 is essentially an approximation of a linear function and standard hyperbolic dependence of bit durability from its rotational speed (more specifically, its upper and lower levels) at a given bit weight:

$$C = T{\bullet}n = Const \quad 3.4$$

This approximation simplified the model and facilitated the determination of optimum levels of the main parameters, within a certain range of bit rotational speed and weight levels, without detriment to the results.

Figure 3–6 shows the dependence of a bit durability in time *T* from its rotational speed *n* at constant bit weight and the average for the number of bit runs in well No. 198 in the Karadag field. Figure 3–7 presents the same dependencies at various axial loads, which indicate a substantial effect of the latter on *T*. [31]

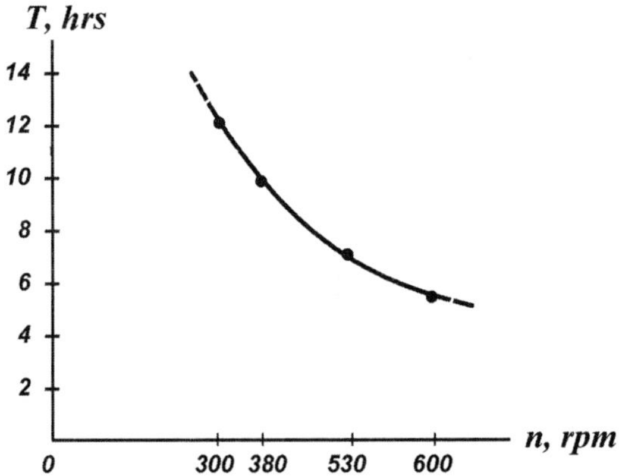

***Fig. 3–6** Dependence of bit durability in time from its rotational speed (at WOB 10–11 t); Karadag field (Azerbaijan), well No. 198, Sabunchinsky and Balakhinsky suites; electrodrills E 215/8, E 215/10, ES 215/2 in combination with frequency converter; bit B11S type*

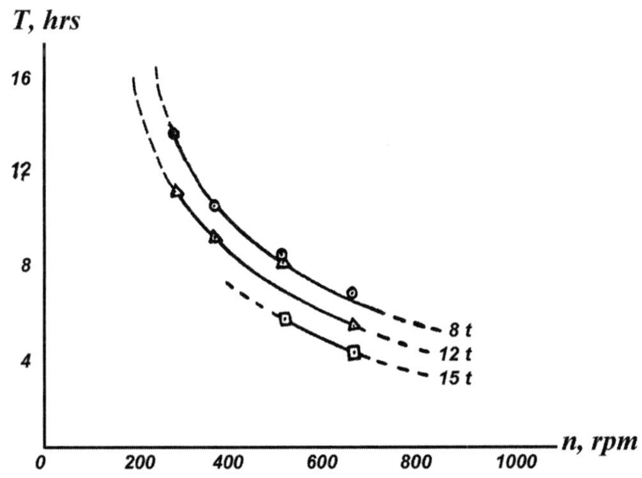

***Fig. 3–7** Dependence of bit durability in time from its rotational speed at various WOB*

Figure 3–8 and 3–9 present the form of Equation 3.3 dependencies that characterize the joint effect of bit rotational speed and weight on *T.* [32]

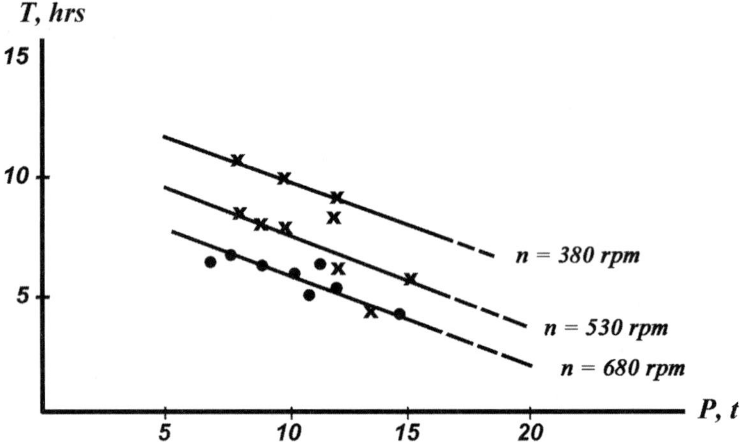

Fig. 3–8 Dependence of a bit durability in time from WOB at various rotational speeds; Karadag field (Azerbaijan), well No. 198, electrodrills E 215/8, E 215/10, ES 215/2 in combination with frequency converter; Sabunchinsky and Balakhinsky suites; bit B11S type

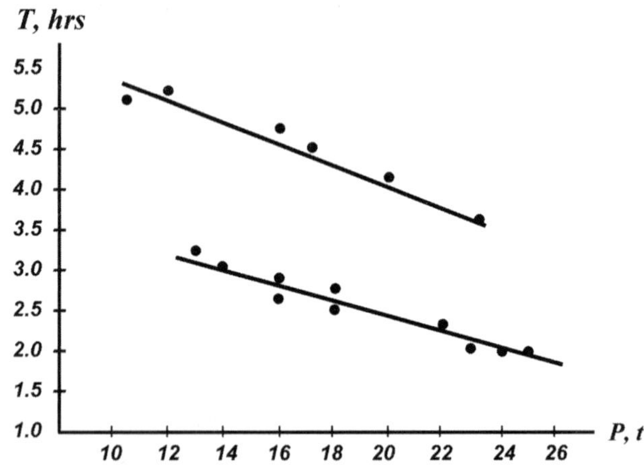

Fig. 3–9 Dependence of bit durability in time from WOB at rotational speed of n=680 rpm; Dmitrovsky field, Samara region, wells No. 168; drilling with electrodrill E 213/8, Interval 557–2494 m

Current penetration rate level in the case of its significant decrease during a bit run is determined from the equation:[33]

$$V_m = Ae^{\frac{-Bt_k}{T}} \cdot n^{\alpha} \cdot P^{m} \qquad 3.5$$

where

A, B are constants dependent on the properties of the drilled rock, bit type, and borehole cleaning conditions

t_k is the overall bit on bottom time

T is the durability of bit bearings

In the technique developed in 1976,[34] a polynomial quadratic model was used for the current penetration rate level.

$$V_m = K \cdot P^m \cdot n^{\alpha}(1 + a_1 t + a_2 t^2) \qquad 3.6$$

where

t is the running time in hours

To determine P_{opt} and n_{opt}, the researchers used a matrix experiment design based on four carefully prepared bit runs.

In the most frequent cases, when the tri-cone bit bearings that determine the bit time on bottom are worn out faster than the cutting structure, corresponding dependencies were made and analyzed for their extremum to determine optimum levels of bit weight and rotational speed to ensure maximum penetration per bit run and bit run speed.[35]

When Equations 3.2 and 3.3 are used, penetration per bit H takes the following form:

$$H = V_{mav} \cdot T = k \cdot n^{\alpha} p^{m}(T_0 - ap - an) \qquad 3.7$$

Bit run speed V_r:

$$V_r = \frac{V_{mav} \cdot T}{T_{trip} + T} = \frac{k \cdot p^{m} \cdot n^{\alpha}(T_0 - ap - bn)}{T_{trip} + T_0 - ap - bn} \qquad 3.8$$

where

$T_{trip.}$ is the trip time of one bit run, including time for related auxiliary operations

By studying the extremum in Equations 3.7 and 3.8, equations for P_{opt} and n_{opt} were developed to ensure the achievement of maximum levels of H and V_r. These equations have the following forms:

For the maximum penetration per bit achievement:

a) One of the parameters fixed:

$$P_{opt} = \frac{m(T_0 - bn)}{a(1+m)} \quad 3.9$$

and in the majority of cases at m=1, Equation 3.9 takes the following form:

$$P_{opt} = \frac{T_0 - bn\bullet}{2a}$$
$$n_{opt} = \frac{\alpha\,(T_0 - aP)}{b(1+\alpha)} \quad 3.10$$

b) for both parameters controlled:

$$P_{opt} = \frac{mT_0}{a(1+m+\alpha)} \quad 3.11$$

and at m=1 will have the form:

$$P_{opt} = \frac{T_0}{a(2+\alpha)}$$
$$n_{opt} = \frac{\alpha(T_0 - aP)}{b(1+\alpha)} \quad 3.12$$

For the maximum bit run speed achievement:

a) one of the parameters is fixed:

$$P_{opt} = \frac{1}{2ma}\left[T_{trip}(1+m)+2m(T_0 - bn) - \sqrt{T^2_{trip}(m+1)^2 + 4mT_{trip}(T_0 - bn)}\right] \quad 3.13$$

$$n_{opt} = \frac{1}{2ab}\left[T_{trip}(1+\alpha)+2\alpha(\grave{O}_0 - bP) - \sqrt{T_{trip}(\alpha+1)^2 + 4\alpha T_{trip}(T_0 - bP)}\right] \quad 3.14$$

b) both parameters controlled:

$$P_{opt} = \frac{m}{2a(m+\alpha)}\left[2T_0 + T_{trip}\left(1+\frac{1-\sqrt{4\frac{T_0}{T_{trip}}(m+\alpha)+(m+\alpha+1)^2}}{m+\alpha}\right)\right] \quad 3.15$$

$$n_{opt} = \frac{\alpha}{2b(m+\alpha)}\left[2T_0 + T_{trip}\left(1+\frac{1-\sqrt{4\frac{T_0}{T_{trip}}(m+\alpha)+(m+\alpha+1)^2}}{m+\alpha}\right)\right] \quad 3.16$$

The optimum values of the drilling process parameters are calculated using Equations 3.9–3.16 based on the coefficient levels of the basic functions, which were determined while drilling the KTW. The further calculations on the Equations 3.7 and 3.8 by substitution of the optimal parameters n_{opt} and P_{opt} allowed us to forecast maximum penetration per bit H and bit run speed V_r levels for new wells drilled in the same and adjacent fields that had similar geological conditions and to achieve maximum borehole deepening rates while developing well drilling programs.

Figure 3–10 shows an example of the utilization of Equations 3.9 and 3.10 in the Stavropol region. Since the study did not address the case of bit failure due to wear of cutting structures when $t_k<T$,[3] it should be said that if calculations for this case are to be made, Equations 3.7 and 3.8 must be analyzed for the extremums of the three variables: P, n and t_k. The levels of P_{opt} and n_{opt} obtained from these calculations by and large exceeded the practical allowable levels. Therefore, in the case of wear on the bit cutting structure is predominant, the maximum practical magnitude allowable for this bit size using the drilling parameters corresponds to the maximum penetration per bit and bit run speed.

Analysis of the effect of round trip speed on levels of the main regulated drilling parameters—bit weight and rotational speed

Equations 3.13–3.16 include T_{trip} trip time per bit run as a constant component, which includes time for related auxiliary operations. The advancement of drilling

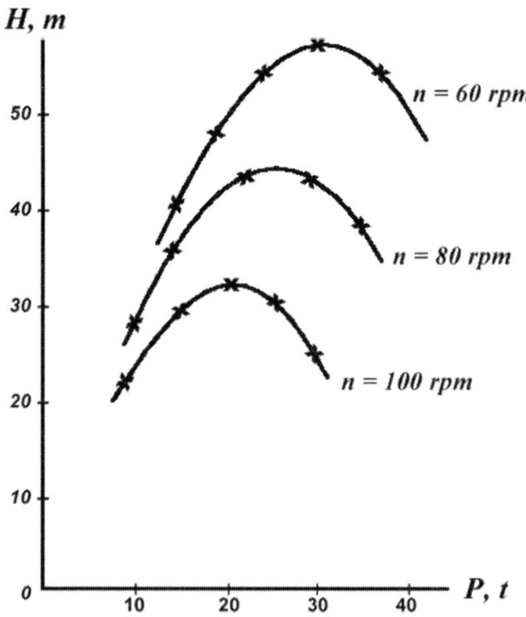

Fig. 3–10 Dependence of penetration per bit 2K214SG from WOB and rotational speed; Mirnensky field, Stavropol region; interval 2050–2300

technologies involved significant amounts of scientific and research work aimed at reducing T_{trip}. The researchers were interested in the effect of this parameter decrease on the main regulated parameters of the drilling process, P and n.

It should be mentioned that measures like the following allowed average round trip time reduction by 50 to 60% for wells with depths of 3000–3200 m:

- increase of a drilling rig drive capacity
- use of aluminum DPs
- increase of drilling rig height to 53 m and joint length to 37 m
- automation of round trip operations using a special tool handling system

Even more significant reductions (5–6 and more times) of trip time and time for related auxiliary operations were achieved by using retractable bits, which do not require DP round trips to replace a bit (*see* Chapter 7 in Volume 2 for details). This method quite distinctly demonstrates the effects of this factor on P_{opt} and n_{opt}. The issue has been studied and the corresponding reports published. The studies were based on information from test runs of retractable two-cone bits in the Saratov region [36] and in the Karadag field [37] using data obtained from drilling well No. 198 with an electrodrill, which was mentioned earlier. Slightly simplified forms of Equations 3.13 and 3.14 were used for building diagrams of Functions $P_{opt}=f_3(T_{trip,}\ n)$ and $n_{opt}=f_4(T_6,\ P)$ with exponents m and α equal to 1.

In this case they take the following forms:

$$P_{opt} = \frac{1}{a}\left[T_{trip} + T_0 - an - \sqrt{T^2_{trip} + T_{trip}(T_0 - bn)}\right] \quad 3.17$$

$$n_{opt} = \frac{1}{ab}\left[T_{trip} + T_0 - bP - \sqrt{T^2_{trip} + T_{trip}(T_0 - bP)}\right] \quad 3.18$$

In the conditions of the Karadag field, T_0 for 269-mm bits was 23–24 hours, whereas a=0.45 and $б$=0.02. [38]

Diagrams of these functions are shown in Figures 3–11 and 3–12.[39]

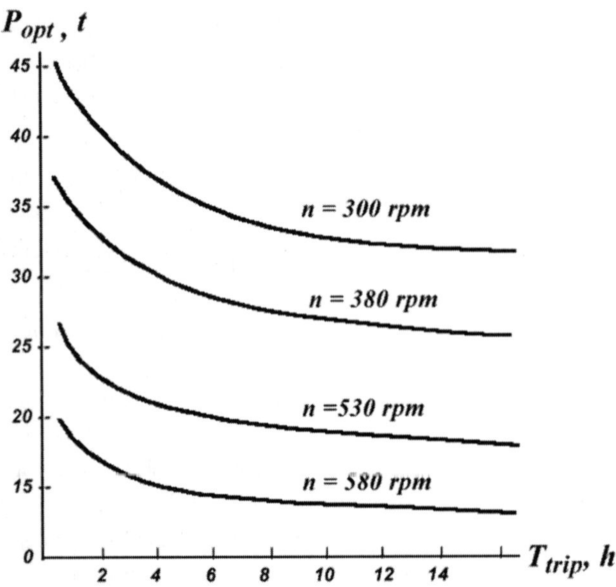

Fig. 3–11 Dependence of optimal WOB from trip time at various rotational speeds

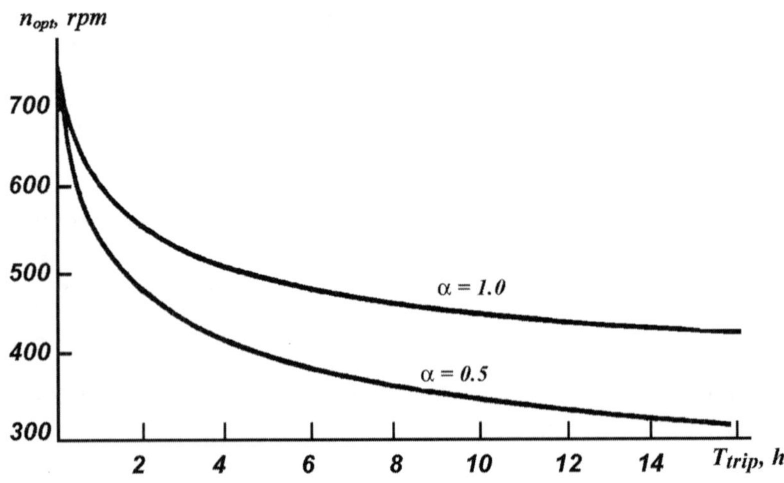

Fig. 3–12 Dependence of optimal rotational speed from bit trip time at WOB 20

Figure 3–11 shows a graphical representation of the function $P_{opt}=f_3(T_{trip}, n)$ indicating that at various n, the level of P_{opt} went up when T_{trip} decreased, especially at $T_{cn6} < 4$ to 6 hours. Furthermore, the rate of this increase was more substantial at lower n.

Figure 3–12 presents $n_{opt}=f_n(T_{trip}, P)$ the curve for the exponent $\alpha=1$ and $\alpha=0.5$. The curve indicates that at very low T_{trip}, the difference of n_{opt} for these levels of α showed a substantial decrease, which meant less significant effect of rock physical and mechanical characteristics on n_{opt} levels for these conditions.

Figures 3–13, 3–14, and 3–15 show bit run speed versus P_{opt} and n_{opt} curves for various T_{trip}.

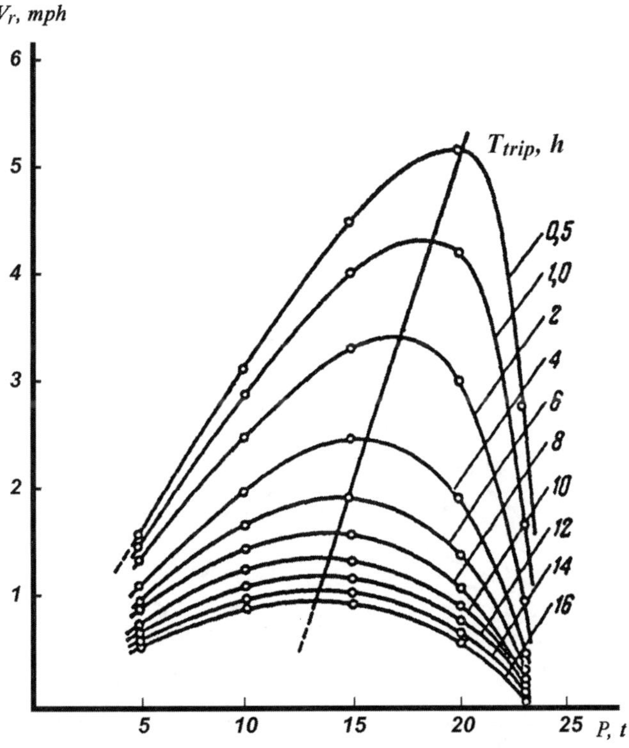

Fig. 3–13 Dependence of bit run speed from WOB at different bit trip times at rotational speed of 680 rpm

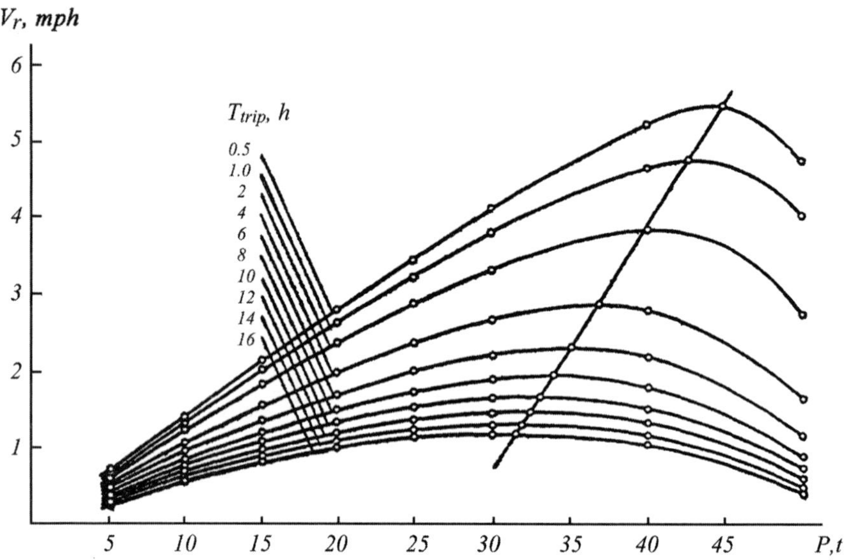

Fig. 3–14 Dependence of bit run speed from WOB at different bit trip times at rotational speed 300 rpm

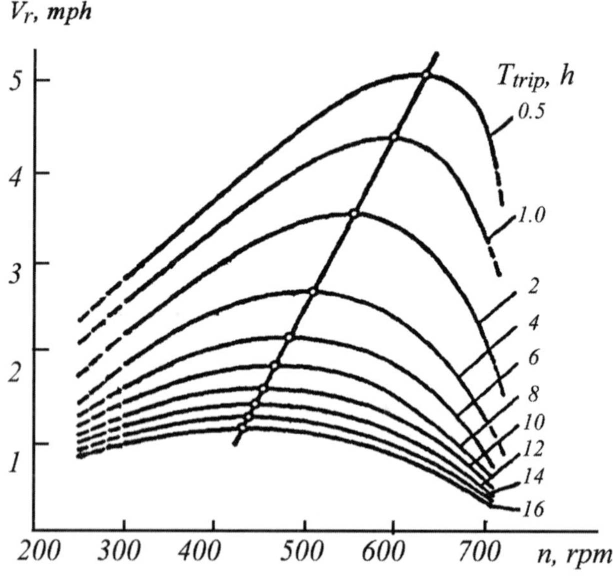

Fig. 3–15 Dependence of bit run speed from rotational speed at different bit trip times at WOB 20 tons and coefficient $\alpha=1$

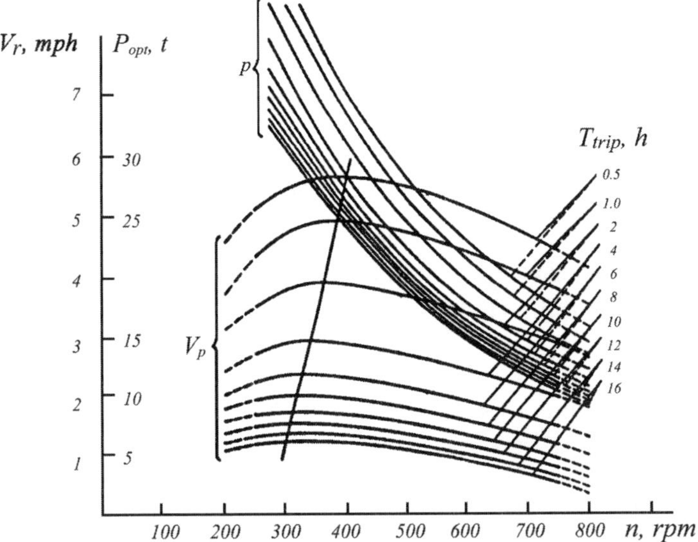

Fig. 3–16 Dependence of bit run speed and optimal WOB from rotational speed at various bit trip times

Figure 3–16 presents a chart, built using the functions shown, which allowed reaching $V_{r\,max}$ with the possibility to change both, P_{opt} and n_{opt} at various T_{trip}.

The $V_r = f(n)$ curve peaks will correspond to the absolute maximum levels of bit run speed for the given drilling conditions and given T_{trip}. This chart was built for the drilling conditions in the Karadag field for the intervals drilled with tri-cone bits.

For example, the chart indicates that for $T_{trip} \approx 2$ hours, the absolute maximum level of the bit run speed of approximately 4 m/hr was achieved at a bit rotational speed of 360 rpm and a bit weight of 34 tons. The axial load was determined by restoring a perpendicular line from the $V_{r\text{-}max}$ point to the point of crossing with the $P_{opt} = f(n)$ curve, which corresponded to T_{trip}=2 hours. The dependencies of the optimum drilling process parameters that were established from the time required for a drillbit change were true for the specific drilling conditions in the Karadag field. Obviously the accepted technique of the test data processing and the following analysis could be used for other drilling conditions.

Well-deepening optimization techniques when drilling with blade-type drag bits

It is well known that rocks with high plasticity (potter's clay and sandy shale) and viscosity are best drilled using blade type bits that apply a dragging action on the formation. The actual data confirming this statement is presented later in this section.

These types of rocks occurred widely in the southern regions of the FSU, particularly in Azerbaijan, where most of the sections were composed of Tertiary and Quaternary Cenozoic deposits. Therefore, blade bits can be applied efficiently in these regions.

According to L. A. Schreiner's classification, the measured hardness of the high-plastic and elasto-plastic rock was 20–50 kg/mm^2 using the flat bottom cylinder die indentation method and the plasticity factor (κ_{pf}) was infinity to 1.8–1.2. The program of drilling test wells in Azerbaijan included test runs of blade bits in wells drilled in the Karadag field and the adjacent Pirsagat field to evaluate their performance in various intervals of a cross-section and to determine optimum drilling parameters. Since this type of bit design did not include bearings, their on-bottom time was limited only by the cutting structure wear and by bit gauge loss. The bit gage loss resulted in reduced penetration rate and gage loss of a borehole and also necessitated reaming.

Similar to the tri-cone bits, the solution in this case could be found by determining basic dependencies of the penetration rate from the main parameters P and n, bit wear in time, and power characteristics. The test runs were made in the Surakhansky and the Sabunchinsky suites of the Karadag field and in the entire cross-section of the Pirsagat field while drilling wells using the rotary and electrodrilling methods. [40] [41]

During the test studies, the researchers determined the following dependencies:

$$V_m(P),\ W_b(P),\ M(P), \text{ and } A_{sp}(P) \text{ at } n=\text{const}$$

$$V_m(n),\ V_m(t), \text{ and } W_b(t) \text{ at } P=\text{const}$$

where

W_b is the bit operating power

A_{sp} is the specific power consumed per drilled unit volume of formation

t is the bit on bottom time

Test bit runs were made in the interval 200–2300 m using 3L and 3LG three-blade jet nozzle bits of 394 mm diameter, whereas 3LG and IRG scraping-cutting type jet bits of 269 mm diameter were run in the interval 2300–3000 m. Figures 3–17 and 3–18 show the designs of these bits. [42]

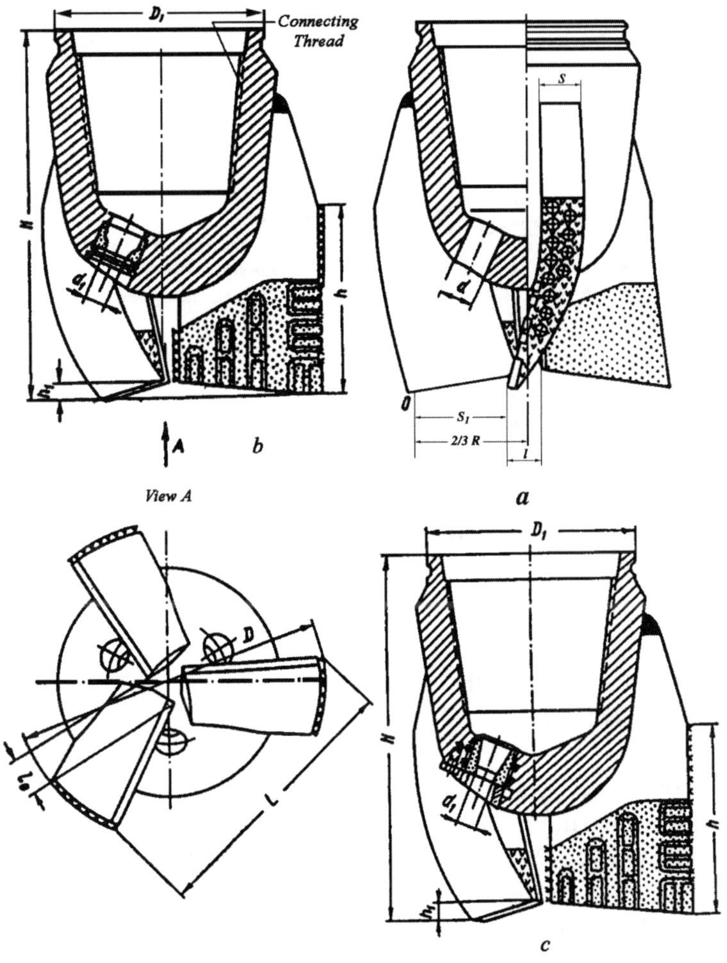

Fig. 3–17 Three-blade bit type 3L and 3LG a) with conventional circulation type 3L b) with jet nozzle type 3LG, designed by VNIIBT c) with jet nozzle type 3LG, designed by AzINMASh

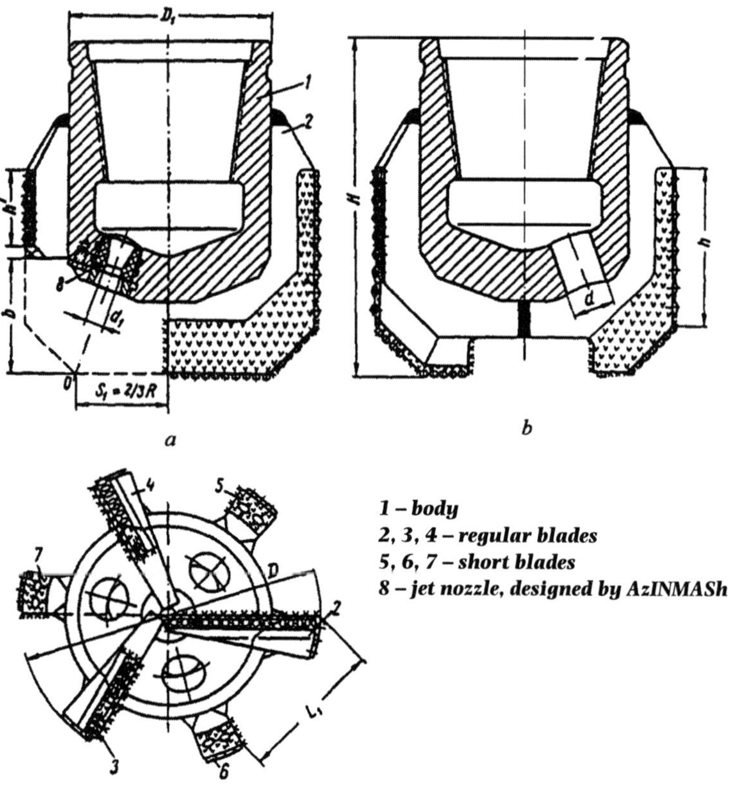

***Fig. 3–18** Multi-blade scraping-cutting-type bits types IR and IRG a) with jet nozzle type IR, designed by AzINMASh b) with conventional circulation (IR)*

The 3LG bits were used with jet nozzles designed by engineers from the Oil Industry Engineering Institute of Azerbaijan (AzINMASh).

The researchers determined dependencies of penetration rate and power characteristics from the drilling parameters by step changing bit weight at various bit rotational speed.

A number of bit runs were made at constant drilling parameters to determine a wear pattern for the blade bits. Figure 3–19 shows the functions that were determined using the technique previously described.

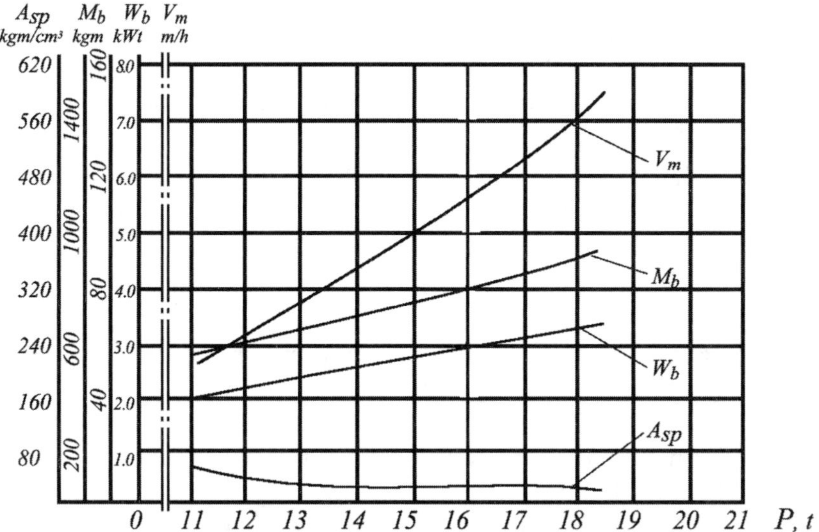

Fig. 3–19 Different drilling parameters dependencies on WOB, Pirsagat field, bit 3LG-394, drilling interval 1100–1650 m, Q=50 l/sec, n=70 rpm; d=16 mm, clay formations

The established function $V_m(P)$ was approximated by the equation:

$$V_m = kP^{\beta} \tag{3.19}$$

where

k is the proportionality factor, depending on the physical and mechanical characteristics of the rock, drilling fluid type or its circulation rate, and bit size

β is the exponent of P, depending on the physical and mechanical characteristics of the rock, drilling parameters, and bit design

Unlike the exponent m for tri-cone bits, the exponent in this case may exceed 1 and, for example, vary between 1.2 and 2.08 for given conditions. A high level of β is explained by the fact that the drag bits realized the most effective way for the rock volumetric failure and bottomhole cleaning.

The conclusion was that while drilling with blade bits, the level of P should be increased to the maximum possible limit restricted only by the operational parameters of the equipment, bit and DS strength, and an acceptable level of the well borehole curvature. This is the drilling parameter's optimization for conditions of the bit cutting structure wear that was described earlier in this chapter when analyzing the mathematical model for the tri-cone bits.

Function $V_m(n)$, at various levels of P, found in the course of the tests and studies are shown in Figure 3–20.

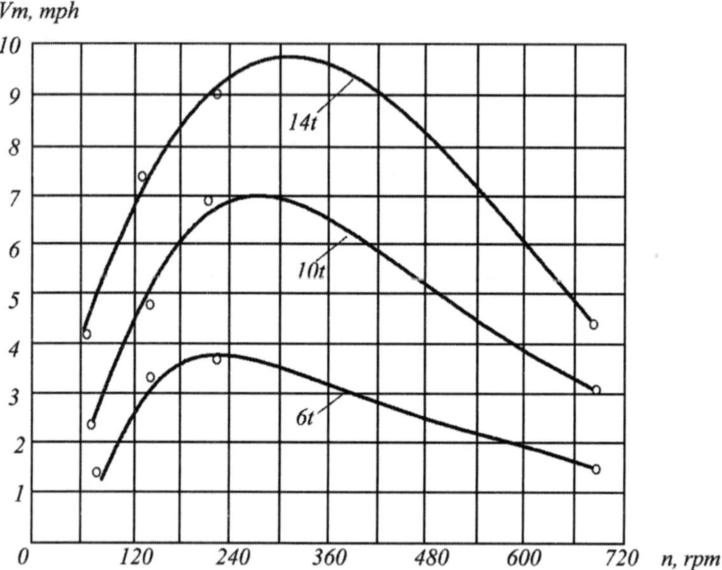

Fig. 3–20 Rate of penetration dependencies on rotational speed at various WOB values, Karadag field, Surakhansky suite, bit 3LG-394, drilling interval 1600–1850 m

This function is described by the following equation:

$$V_m = A_2 n e^{-\theta n} \quad 3.20$$

where

A_2, θ are empirically found coefficients, depending on bit design, drilling parameters, rock physical and mechanical characteristics, and type of drilling fluid

Similar to the tri-cone bits model, these coefficients were determined by taking the logarithm of this equation after obtaining two actual levels of V_m at two different levels of n, all other conditions being equal.

The ROP reduction in time, when using blade type bits, occurred exclusively due to the bit cutting structure wear, i.e., change of the bit contact surface area.

Figures 3–21, 3–22, 3–23, and 3–24 show functions $V_m(t)$ determined from tests. The diagrams indicate that the drilling penetration rate during a bit run, when using blade type bits, went up from the original level to the maximum and then decreased smoothly. This fact could be explained as follows. Penetration rate increase occurred as a result of the bit contact surface area reduction during break-in because of the hard facing exposure and the ridge type surface formation. Later on as the cutting surface was worn out, the area of bit contact with a borehole increased, but the area load levels went down, which led to a lower penetration rate.

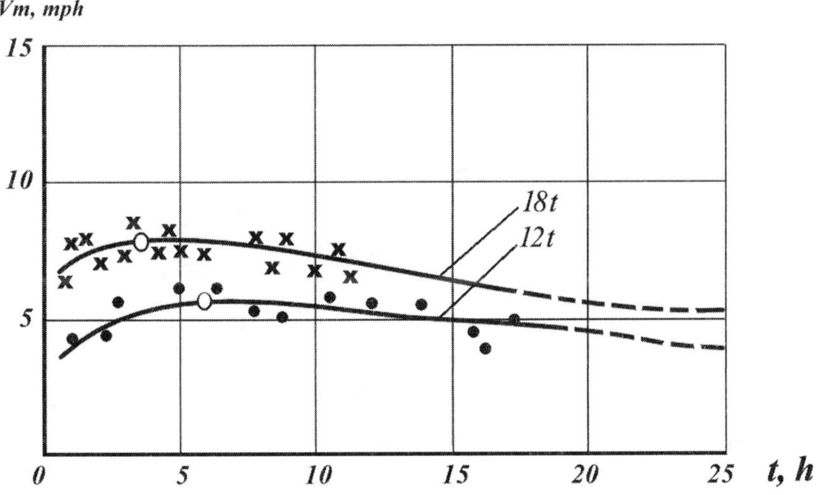

***Fig. 3–21** Rate of penetration dependencies on drilling time, Karadag field, Surakhansky suite, drilling interval 1600–1850 m, bit 3L-394, n=70 rpm*

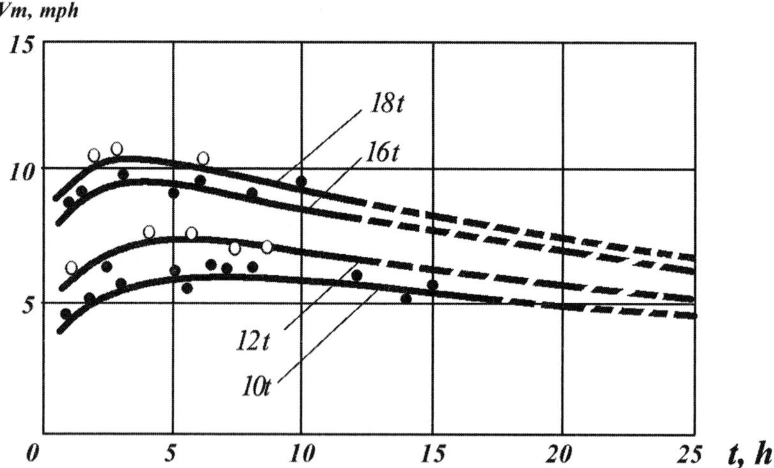

Fig. 3–22 Same as Fig. 3–21, but with n=140 rpm

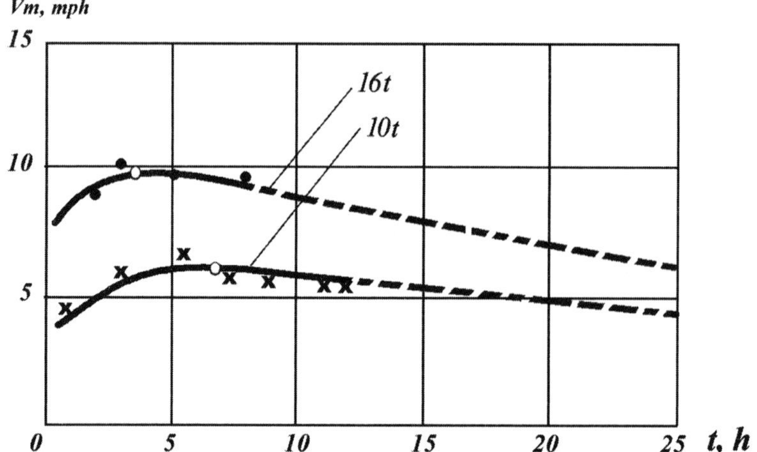

Fig. 3–23 Same as Fig. 3–21, but with n=220 rpm

The function $V_m(t)$ was approximated using the exponential function:

$$V_m = Ae^{-k_1 t} - Be^{-k_2 t} \quad 3.21$$

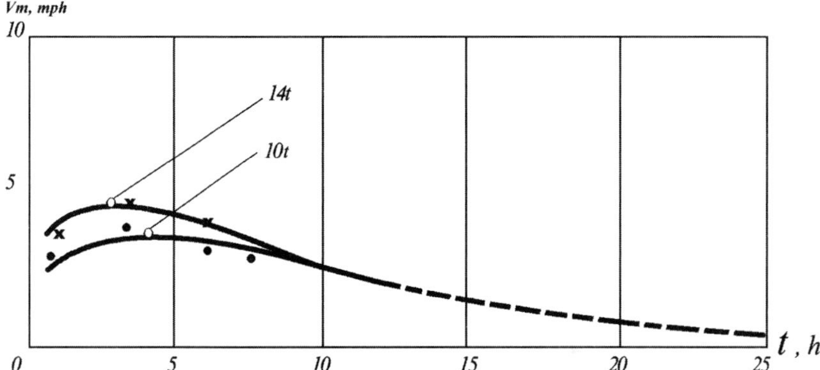

Fig. 3–24 Same as Fig. 3–21, but with n=680 rpm

where

- t is the bit on bottom time
- e is the natural logarithm base
- A, k_1, k_2 are coefficients, dependent on rock physical and mechanical characteristics, type of drilling fluid, drilling process parameters, and bit design size
- B is the coefficient for designed constant rotational speed, depending on rock physical and mechanical characteristics, and bit design and size

The latter coefficient accounts for a specific pattern of change in the form of the bit blade to a tapered form (wear starts at the gage row which has the fastest peripheral speed and continues toward the middle).

The diagrams in Figures 3–21, 3–22, 3–23, and 3–24 indicate that an increase in P resulted in faster bit wear. In the area located to the right of the maximum penetration rate level, the absolute values of the function $Be^{-k_2 t}$ were so small they did not affect the pattern of V_m change for practical purposes. Therefore, Equation 3.21 at high P values can be reduced to the following form:

$$V_M = Ae^{-k_2 t} = V_0 e^{-k_1 t} \qquad 3.22$$

where

- k_1 is the decrement of drilling speed decrease due to bit wear
- V_0 is the initial drilling speed

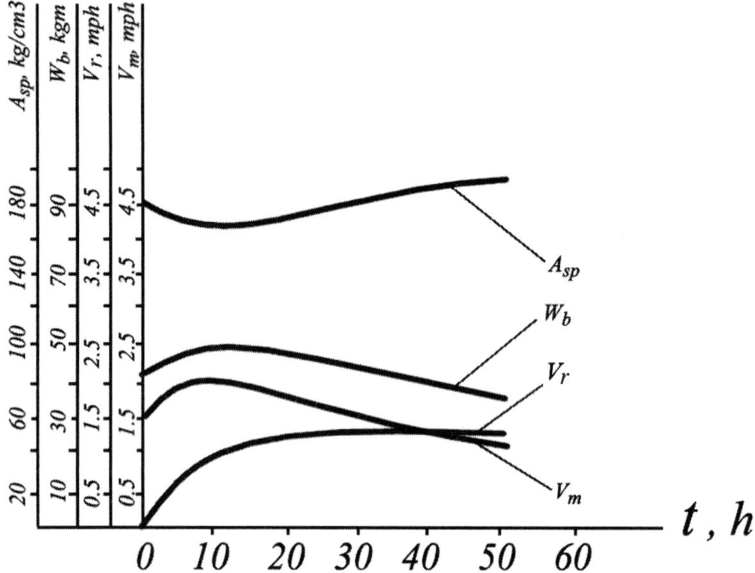

Fig. 3–25 Drilling parameters dependencies on drilling time, Pirsagat field, drilling interval 2600–3000 m, bit 3LG-269

This equation complies with the currently accepted exponential law of speed change during a bit run.

The diagram in Figure 3–25 shows power variation while drilling with blade bits.[43]

Functions $W(t)$ (curve 3) and A_{sp} (curve 4) indicate that before the maximum penetration rate, curve 1 was achieved. Power W required for rock destruction was increasing, whereas specific energy consumption per drilled out rock unit volume A_{sp}, was going down.

The maximum power used and the minimum energy rate corresponded to the maximum V_m level. With the decrease of V_m, W went down and A_{sp} went up.

This identified the principal basic functions of controlled drilling parameters when drilling with blade bits, [44] as well as numerical values of the basic functions'

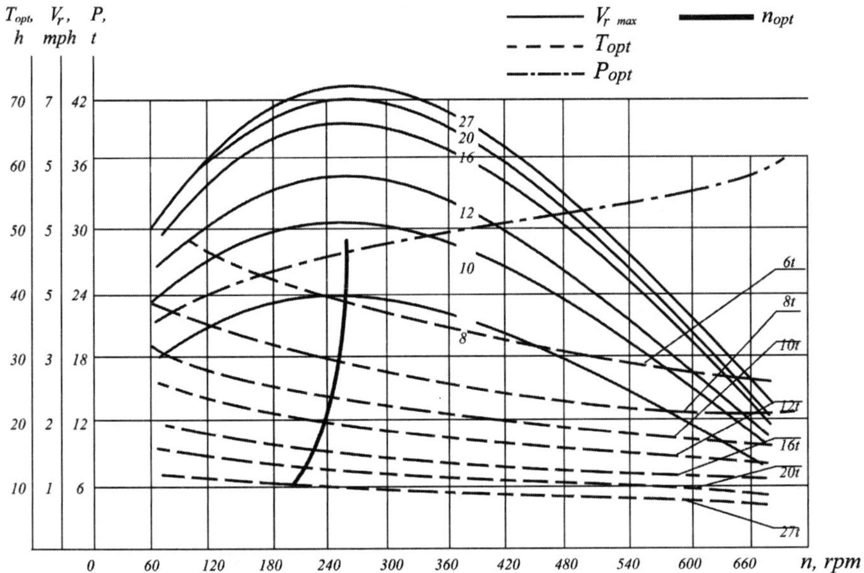

Fig. 3–26 Optimal drilling parameters dependency on the bit RPM

coefficients, that were determined using a special technique, [45] and allowed building the nomograph shown in Figure 3–26. [46] The chart made it possible to determine P_{opt} and n_{opt} values to find $V_{r\text{-}max}$ and to account for the optimum bit-on-bottom time *t*. The nomograph also helped determine values of controlled parameters in the cases of one of the parameters' level limitations. For example, this was done when *P* was limited due to certain components of the system being overloaded, which might result in reduction of DS strength or in possible curving of the well borehole.

To summarize optimization of the drilling process when drilling with blade bits, the use of this type rock bit has seen a significant decrease in the last decades. These bits were unreasonably committed to oblivion and are seldom used now. At the same time, the available information indicates that when used in certain geological conditions, the bits showed excellent results that could hardly have been accomplished when running other types of bits.

Results of Drilling Experimental and KTWs and Application of the KTW Technique in Developing Certain Fields

Wells Drilled in Azerbaijan [47]

The first experimental wells drilled in Azerbaijan, Samara province, and Bashkiriya in 1961 and 1962 were constructed with planned test activities and goals aimed at obtaining information on rotary, turbodrilling, and electrodrilling technology.

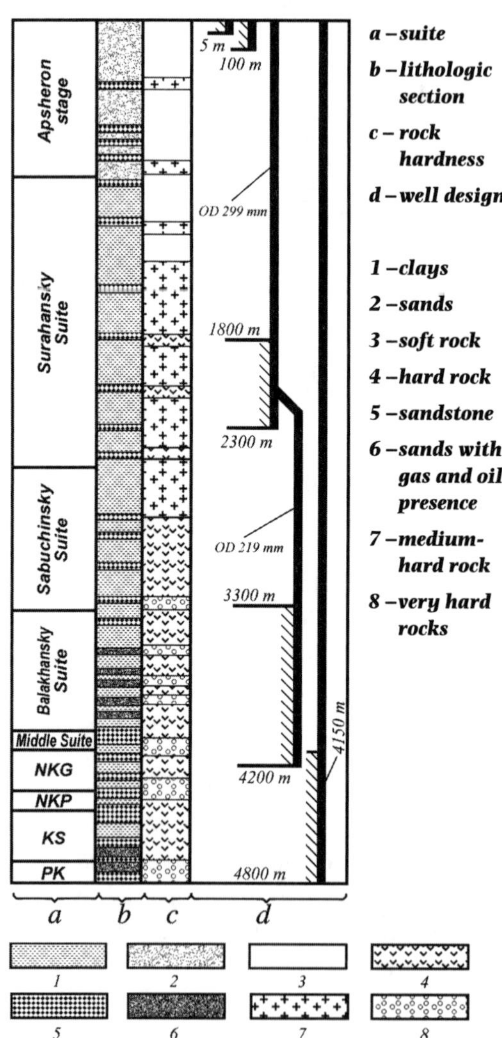

Fig. 3–27 Lithology section and design of test wells No. 156, 153, and 157

The program of drilling test wells in Azerbaijan was developed by the VNIIBT in cooperation with the scientific and research institute AzNIIburneft and the Azneft production company from Baku.

From 1961 through 1964, three wells were drilled in the Karadag-Damba area using various drilling methods: No. 156 (rotary), No. 153 (turbodrilling), and No. 157 (electrodrilling), with TDs of 4800 m.

Figure 3–27 [48] presents a geological cross-section of the Karadag-Damba field along with the test well design.

Drilling conditions in the southeastern part of the Karadag structure, where the test wells were drilled, were quite complicated because of gas shows, hole sloughing, tight sections, and formation arching, which led to DP dragging, sticking, and keyseating problems.

Besides the standard measuring and control equipment, the following instruments were installed on the rig to enable control and analysis of the results of the drilling process: the KPB-3 driller's control console, recorders (ammeter, voltmeter, wattmeter), and the RGR-4 flow meter. These instruments allowed continuous registration on chart strips of penetration rate, bit weight, pump discharge operating pressure, power input, amperage, and voltage. Drilling the test well involved using the Uralmash 43-61 drilling rigs with three U8-4 pumps, two shale shaker-and-conveyor combination units, the AKB-3 DP tongs, the PKR-Sch8 pipe slips, the PIRSh-4 rotary drive. In addition, the rig that drilled well No. 157 (electrodrilling) included such equipment as drillbit feed control mechanism AVE-1 and a frequency converter unit that enabled control of the electrodrill shaft rotational speed by bringing the electrical frequency down from 50 Hz to 34.5 Hz.

Well No. 156 was drilled using DRCs with diameters of 273 mm, 203 mm, and 178 mm. Total length of the DRC section was from 140 m to 240 m, which made possible an axial load increase up to 25 tons. While drilling this well, the drillers used such tools as jet nozzle bits (both three-blade and tri-cone bits) and higher power-consuming M and MS-type cone bits instead of the commonly used bits with S-type cutting structure. They also used strength category E steel DPs, and the most effective chemical (at that time) for mud treatment—polyphenol wood-chemical (PFLK).

Use of the specially selected equipment and optimized drilling and mud circulation parameters when drilling well No. 156, spud-in December 1961, achieved much better results compared to well No. 166 drilled earlier in the Karadag-Damba field using the rotary drilling method (Table 3–1). At some intervals, the drilling results were even more impressive.

TABLE 3–1
Experimental Well No. 156 Drilled in Azerbaijan at Karadag-Damba Field in Compared with Best Well No. 166 Drilled Before, Rotary Method Used in Both Cases

Well No.	Drilled interval	Number of bit runs	Average penetration per bit, m	Penetration rate, m/hr	Bit run speed, m/hr
156	50–3,852	78	50.5	4.25	2.18
166	110–3,898	99	34.5	2.42	1.45

Note: The drilling results are compared only for the interval drilled to the setting depth of the 219-mm string. A serious downhole failure occurred in well No. 156 at 4071 m related to a pipe being stuck, which resulted in drilling a second hole that also never reached the TD. For technical reasons, both holes were plugged and abandoned.

A representative example is the interval that was drilled in the Apsheron-Akchagal suite deposits from below the 426-mm conductor casing in which three-blade bits were run. During the first run, a 14-mm jet bit with ceramic nozzles penetrated 944 m with an average penetration rate of 87.2 m/hr, i.e., the entire interval was drilled in 10.8 hours. This allowed drilling a 1000-m interval in approximately 15–18 hours. This was accomplished by providing the required bit weight level of up to 20–25 tons, jet nozzle outflow velocity up to 110 m/sec, and a DS rotational speed up to 220 rpm. Not one of the wells that had been drilled previously exceeded 15 m/hr with the turbodrilling method or 34 m/hr with the rotary method.

The unique results from drilling in Apsheron-Akchagal deposits allowed a conclusion about the high potential performance results and the application of low cost of jet nozzle blade bits in this and similar deposits at deeper intervals. This conclusion was enhanced by the possibility of achieving even better results by increasing bit weight, improving borehole cleaning with higher mud circulation rate, and by using a drillbit feed control device. Therefore, the facts encouraged promoting wide use of these bits.

The analysis of the data, along with simple calculations, indicated that just 9 bits (Rather than 15–17) could drill the interval for a 299-mm intermediate casing in 120–130 hours at an average penetration rate of 18–24 m/hr. By excluding the downtime, reducing the time of round trips and mud pump repair, and reducing downhole failures such as DP sticking, drillers could achieve an overall drilling speed of 5000–6000 m/rig-month while drilling the interval for the intermediate casing. This meant that the interval could be drilled and cased in only two weeks.

The B11MGL bits and jet nozzle 269 mm diameter blade type bits showed excellent results when drilling below the 11-in. (299-mm) intermediate string. Comparisons between the best results from wells drilled using turbodrills in this interval of the Balakhansky suite and the "medium series" (i.e., to the depth of the second intermediate string) showed the advantage of using the rotary drilling method when running high power-consuming tri-cone jet nozzle bits with long teeth.

These type bits were also run in combination with an electrodrill in well No. 157 that was spud-in on March 7, 1962. In this well, however, blade bits drilled much smaller intervals, whereas a larger number of the B11MGL and B11MSG type tri-cone bits with high power consumption were used. In some sections of the borehole, the bits were used in combination with two-section electrodrills. High mud pump pressure and forced drilling practices were also used while drilling this well.

The performance results from the well were better than any of the other electrodrilled wells, but they were not as good as the results from drilling well No. 156 using the rotary method. Table 3–2 shows data from the wells drilled using the electrodrill method.

TABLE 3–2
Drilling in Azerbaijan at Karadag-Damba Field with Electrodrills: Experimental well No. 157 Compared with Previous well No. 198

Well No.	Drilled interval	Number of bit runs	Average penetration per bit, m	Penetration rate, m/hr	Bit run speed, m/hr
157	97–4,000	124	29.2	4.05	1.75
198	100–4,003	171	22.2	2.96	1.26

The electrodrill process experienced many difficulties because of low durability of the electrical cables and the electrodrilling motors themselves at that time. A total of more than 40 cable faults were revealed in various sections of the cable and motor. As a result, 10 bit runs were not completed. This fact significantly affected the general performance results of electrodrills.

While drilling the interval below the first intermediate string, the drillers could use a frequency control unit, which allowed them to reduce the speed of the high-torque two-section electrodrills from 680 rpm to 450 rpm. This new technique substantially improved drilling results even though the B11S non-jet nozzle bits with medium tooth length were run instead of the B11MGL and B11PSG bits, the latter being unavailable. These results are shown in Table 3–3, [49] which indicates that due to a high bit weight of 20–25 tons and good borehole cleaning (Q=31–35 liters/sec), drillers were able to increase the penetration rate and the penetration per bit by 2–2.5 times.

TABLE 3–3
Comparison of Drilling Results from Well No. 157, Drilled with Standard "E" and sectional "ES" Electrodrills

Interval, m	Type of electro-drill	Nos. of bit runs	Average penetra-tion per bit, m	Average time of a bit run, hr	Average ROP, m/hr	Bit type	Drilling parameters			
							WOB	rpm	Pressure drop, kg/sec • m²	Current, ampere
2,399–2,508	E215/8	5	21.8	9.05	2.41	B11S	14–20	450–680	125	140
2,508–2,550	ES215/2	1	42.0	8.50	4.95	B11S	20–23	450	125	140
2,550–2,597	E215/8	2	23.5	7.10	3.31	B11S	16–20	450	125	140
2,597–2,639	ES215/2	1	42.0	6.25	6.62	B11S	18–22	450	130	110–120

Increase of the well depth to 4000 m and deeper while drilling in the "medium series," frequent cable faults, and a 30–40% reduction rate of circulation fluid due to large hydraulic losses significantly worsened the drilling results to the point where further well deepening became unfeasible. For that reason, drilling was stopped at the depth previously noted. Next, the production string was run in to enable production from horizon VIII of the medium series.

The experience from drilling this well gave an impetus to further development and improvement of the entire electrodrilling method. This development and improvement work was instrumental in eliminating the noted drawbacks and achieving very good results using the electrodrills for deep well drilling. The technology of electrodrill sectioning and frequency control units have not found wide application since reduction gear inserts technology proved to be simpler and more reliable in achieving higher bit torque and lower rotational speed (*see* Chapter 2 for more details).

The drilling of well No. 153 began in March 1962. It was drilled using the turbodrilling method and was the only well of this kind that was drilled to the TD. Table 3–4 presents drilling results from this well compared to the best wells drilled to the running depth of the second intermediate string using turbodrills.

TABLE 3–4
Experimental Well No. 153 Drilled in Azerbaijan at Karadag-Damba Field Drilled with Turbodrills Compared to Previous Wells

Well No.	Interval, m	Number of bit runs	Average penetration per bit, m	ROP, m/hr	Bit run speed, m/hr
153	62–4,018	140	27.1	5.6	1.95
143	45–4,050	174	22.8	2.4	1.21
144	50–4,000	155	24.9	4.2	1.63

As mentioned in Chapter 2, the three-section turbodrills 3TS5B-9" were tested for the first time in well No. 144 in combination with the B11S bits. This test aimed at evaluating their efficiency prior to running them in well No. 153. In well No. 143, drillers used only the TS5B-9" two-section turbodrills. The 3TS5B-9" turbodrills proved to be quite efficient, which contributed greatly to an increased penetration rate and successful completion of well No. 144.

This type of turbodrill was used in well No. 153 combined with more efficient high-power consumption bits such as types B11MGL and B11MSG. During their running, higher mud pump pressure levels of up to 170–180 kg/cm^2 were used. In certain intervals, lower rotational speeds of 350–435 rpm were employed, which resulted in even better drilling results (*see* Chapter 2 for more details).

The drillers were not able to achieve good commercial results from drilling test wells, because of a number of organizational problems, downhole failures, and the large amount of repair work that was required. However, the excellent technical results and the large amount of research work allowed a number of important conclusions about drilling in the Karadag field and a number of other fields in Azerbaijan with similar drilling conditions that included:

1. Integrated use of properly combined available technology and equipment proved to be one of the powerful methods for improving drilling results and well economics.
2. Rotary drilling with running jet nozzle blade and cone bits with long teeth (*M-type*) combined with certain drilling parameters were the most effective method of drilling to depths of 3700–4000 m.
3. In deep intervals, the results of turbodrilling using diamond bits were better, compared to rotary drilling, even with heavy mud.
4. Design and operational characteristics of certain commercially produced drilling equipment, as well as of some measuring and control equipment and instrumentation, did not meet the requirements of the existing deep well drilling technology.

The experience with drilling the first test wells in the Karadag-Damba field was used for wide-scale development drilling on this and other fields in Azerbaijan (the Alyatsk Ridge field, the Pirsagat field, etc.). This experience was applied most successfully while drilling well No. 57 in the Alyatsk Ridge field of the Prikurian depression. The TD of the well was 5000 m. Drillers used the rotary drilling method and drilling tools such as the jet nozzle blade and tri-cone bits, TBVK140 strength category L and M DP, and balanced DRCs of 254 mm and 203 mm diameter. Bit weight, when drilling to the depth of 4300 m, was up to 27 tons with a rotational speed of 140 rpm compared to 12–14 tons and 70 rpm in the previous wells. As a result, the well drilled in 1967 set a record for overall drilling speed of approximately 500 m/rig-month. The actual well depth was 4900 m.

Table 3–5 presents a comparison of the drilling results from this well with other wells drilled with good results both before and after.

TABLE 3–5
The Drilling Speed Record-holder Deep Well No. 57 in Comparison with Other Wells Drilled in Azerbaijan

Well No.	Well, Depth, m	Number of bit runs	Penetration per run, m	ROP, m/hr	Finished in	Overall drilling speed, m/rig • month
57 the Alyatsk Ridge	4,900	70	70	2.4	1967	500
29 the Alyatsk Ridge	4,911	130	37.8	1.33	1966	331.8
46 Karabagly	5,005	120	41.7	2.03	1965	327
83 Kyurovdag	5,001	207	24	1.92	1965	175.4
30 Kyanizdag	5,022	98.0	51.8	1.41	1966	207
401 Kyurovdag	5,000	178	28	2.11	1967	244.4
427 Kyurovdag	5,021	218	23	2.12	1967	259.2
2 Umbaki	5,000	117	42.7	1.98	1967	276
75 Bulla	5,260	164	31.9	1.52	1968	348.3
61 Karabagly	5,110	128	39.2	2.02	1968	233

The results of drilling well No. 57 indicated a high potential for improving results of deep well drilling in Azerbaijan and some other regions. In the followed years, the experience from drilling well Nos. 153, 156, 157, and 57 was used successfully at such fields as the Pirsagat, the Sangachaly and others that featured similar complicated geological conditions, which required use of weighted drill mud with a density of >2.0 gr/cm^3. Although the overall drilling rate of 500 m/rig-month (i.e., drilling 1 well during a 10-month period) nowadays cannot be considered an outstanding achievement, but more than 30 years ago, it was impressive. All these wells were exploratory. They were drilled in remote regions with no access roads or other infrastructure, which led to a significant amount of downtime because of problems with materials supply, a large amount of repair work, and other reasons.

The purpose of showing this data here is to prove that even with problems, a carefully developed and implemented drilling program achieved relatively good results. This approach would prove even more efficient today.

Wells drilled in the Samara Region [50]

Technique of well drilling using water circulation. Before describing results of the test well drilling in the Samara region, it is important to touch upon one of the most distinctive features of the well drilling technique in the Urals and Volga oil and gas provinces where the Samara region is located. This feature was the utilization of water, rather than clay mud, for cleaning the borehole while drilling.

The first attempts at using water for borehole cleaning were made in 1937 and 1938 in the Tuimazin field in Bashkiriya region. This was a forced measure because of the poor supply of clay and chemicals. At that time, these attempts were not very successful because drillers used rotary drills and low capacity mud pumps that could not provide the required quality of borehole cleaning, which resulted in a number of drilling problems. Still, the very possibility of drilling with water circulation was quite important. As is often the case in the history of science and technology developments around the world, it was merely a coincidence that resulted in launching and developing a new trend of well drilling technology in the Urals and the Volga regions. Combined with the turbodrilling method, this new technology proved very cost efficient. [51]

Besides the obvious advantages of using water rather than clay mud, such as eliminating the cost of clay and chemicals, no transportation requirements, reduced wear rate on mud pump components, and fewer hydraulic pressure losses in the circulation system, the main gain was both a higher penetration rate and higher penetration per bit, other conditions being equal. Furthermore, the increase of the two latter parameters was even higher for turbodrilling compared to rotary drilling.

This can be explained by the decrease of the differential pressure ΔP in the well-and-formation system. When it approached the zero level, it resulted in a linear increase in penetration rate, depending on the bit rotational speed, without a plateau level at high rotational speed as is shown in Figure 3–5. The wide use of water circulation became an important factor in cleaning drill cuttings from the borehole of wells drilled in the Urals and Volga oil provinces.

In 1952, 32,000 m were drilled using water circulation, whereas more than 1,500,000 m were drilled in 1955. In the following years, the amount of drilling that used water circulation increased even more as drillers applied ways of using special additives for the circulating water treatment, which reduced fluid loss in a formation and efficiently drilled even thin clay and marl stringers that were normally destroyed (washed out) by water.

The fact is that the geological section of fields located in these oil provinces was primarily composed of old Paleozoic deposits, primarily carbonaceous rocks that featured a high level of hardness and stability. This last fact was extremely important, because it helped maintain the stability of a well borehole when drilling with water circulation.

We believe that the differential pressure decrease to zero, while drilling through carbonaceous rock with water circulation, was related to a phenomenon studied by the academician P. A. Rebinder, professor K. F. Zhigach, and professor Schreiner that was highlighted in the scientific report they wrote. [52] The report suggested the following explanation of the mechanism of water-affected rock destruction.

During a mechanical failure of a solid body, affected by external forces, transmitted to the body by some tool, an intensive fracturing zone was created in the so-called pre-failure area of the layers, adjacent to the failure surface. The pre-failure zone contains a pattern of wedge-shaped micro-fractures, formed around crystal lattice defects or weak points. When external forces were not applied any more, deep fracture zones became smaller, fracture thickness through the entire depth was getting smaller, so that the fracture acquired primordial form and might even close up.

When a solid body failed in a liquid medium with minimum viscosity, dissolved surfactants penetrated into the system of micro-fractures and slowed down their spontaneous closing after external forces were not applied any more. When cone bits were used for rock destruction, their teeth periodically affected rock by hitting it. Part of the energy of this blow was used to cut pieces out of a rock mass, the remaining energy was used to form a weakened pre-failure zone.

The authors assumed that the presence of an active medium, such as pure water or water with some surfactants, extended the life of the weakened pre-failure zone and, when one impact followed another, the accumulated effect improved the efficiency of bit performance.

We do not deny a certain positive effect of this factor. Yet, we believe the process of rapid equalization of formation pressure with water column hydrostatic pressure was the main benefit in the formation of a pre-failure zone that featured an intensive fracture pattern, i.e., the formation of conditions at which ΔP equaled zero. This provided the prerequisites for the sizeable increase of penetration rate, especially at forced drilling modes with high bit weight and the rotational speed of turbodrills and other types of DHMs.

As previously mentioned, the 1950s witnessed a rapid growth in drilling volumes that used water circulation for borehole cleaning. According to the data from 1959, this technology was used in drilling 93% of the wells in the Tatariya region, 65% in Bashkiriya, 40% in the Samara region, and 50% in the Perm region with more than 2 million meters total footage drilled in these regions alone. A combination of the water borehole cleaning technique with forced drilling practices achieved truly outstanding results in a number of wells.

From 1957 to 1959 the overall drilling rate in Tatariya was 5000–5500 m/rig-month in wells that showed the best drilling results with TD of 1700–1800 m, meaning each well was drilled in 10 days. However, the actual bit-on-bottom time was only 60–70 hours at an average penetration rate of 25–30 m/hr. These results were shown in rocks with a Shreiner hardness of 150–400 kg/mm^2. Furthermore, the hardness of certain chert-bearing limestone and dolomite stringers was much higher than that shown.

Implementation of the water circulation drilling technique faced problems that had to be overcome, such as the stability of drilled clay stringers, complicated well electrical logging due to water salinization when drilling through water-saturated formations, and a number of other problems. Several research institutes carried out work to find solutions for these problems, which allowed developing various additives and chemicals for the water-based mud treatment.

Even though utilization of water as a drilling fluid was quite efficient, the measures taken could not provide a comprehensive solution for the overall well drilling optimization process. Rather, these measures suggested only a partial solution. Therefore, the KTW drilling program aimed at developing a more comprehensive solution for the issue.

Experimental well drilling. Test wells were drilled in the Samara region from 1962 through 1965 under the programs jointly developed by engineers from the VNIIBT institute and specialists from KuibyshevNIIneft and Pervomaiburneft drilling company.

Most of these wells were drilled in the Dmitrov field such as well Nos. 156, 90, and 157 drilled using the rotary method, well Nos. 154, 149, 150, 186, and 250 using the turbodrilling method, and No. 168 using the electrodrilling method. The wells featured a single string design, a conductor casing of 11-in. to 12-in. diameter set at a depth of 300–320 m, and a production string with a diameter of 5-in. or 6-in. run to the depth of 3000 m. The wells were drilled using an Uralmash 4E-61 rigs with two U8-4 mud pumps. The rigs were equipped with a KPB-3 driller's console, as well as with electrical measuring and control instrumentation such as ammeter, voltmeter, and wattmeter. The interval from the surface to the top of the Tula horizon at a depth of 2000–2100 m was drilled using water circulation to clean drill cuttings from the borehole.

DSs run in these wells included 200–280 m of 178 mm diameter DC, 140 mm DP of electromagnetic (EM) strength steel, aluminum DP, and drillbits with a diameter of 214 mm that had various types of cutting structures made by the Samara bit plant. Well No. 156 and well No. 90 were drilled at an operating pressure of 120–160 bar from the mud pumps with a drill mud outflow velocity from bit jet nozzles of 100–140 m/sec. Bit weight varied between 20 tons and 28 tons. Because of steel DP failures and serious lost circulation problems, well No. 156 reached a depth of only 1772 m, and well No. 90 reached 2475 m.

When drilling well No. 157, drillers used the RPD-2A bit feed control mechanism and aluminum DP. The mud pump output rate was 40–52 liters/sec. The bit weight, applied to 214-mm bits, varied in the range of 18–28 tons at a rotational speed of 35–60 rpm. The well was drilled to a TD of 2963 m.

The rotary drilling method had never been used in this area before, and so the results from well No. 157 could not be compared to those from any other wells drilled using the same drilling method. However, it is worth mentioning that the overall drilling rate of 593 m/rig-month exceeded the average results from wells drilled in that region using turbodrills.

Drillers ran the E215-8 electrodrill on 146-mm DPs combined with a frequency regulated unit while drilling well No. 168. The frequency regulator enabled a reduction in rotational speed of the electrodrill from 680 rpm to 340 rpm. In addition, the AVT bit feed control unit was used. Electrical power was supplied to the electrodrill through a two wire cable and a DP that performed the function of the third wire. The power cable sections were equipped with KST-1 connecting plugs.

Because of the low durability of electrodrills in these wells, drillers were not able to achieve high results. However, the research work that was carried out while drilling the wells was of great value.

The rig that drilled well No. 154 had been outfitted with some additional equipment such as the AVE-1 bit feed control unit, the ASP-3 automatic pipe handling unit, and hydro-cyclones for the cuttings removal system. The TS5B-9" type two-section turbodrills were run in the well, keeping the mud pump pressure up to 140–180 kg/cm^2. Drillers ran in 269 mm diameter bits with various types of cutting structures. Use of some enhanced equipment, along with optimized drilling practices, allowed a significant improvement in the drilling results of well No. 154 compared to the wells previously drilled in the Dmitrov field using the turbodrilling method (Table 3–6).

TABLE 3–6
Experimental Well No. 154 Compared to Best Wells Drilled Previously in Samara Region, Dmitrovsky Field, Turbodrilling

Well No.	Well, Depth, m	Bit runs	Penetration per bit, m	ROP, m/hr	Bit run speed, m/hr	Overall drilling rate, m/rig • month
154	2,967	169	17.6	8.2	2.45	972
Wells with best drilling results						
306	2,955	215	13.8	7.26	1.93	744
344	2,950	215	13.7	6.74	1.90	789
Average results						
For 11 wells	2,959	212	14.0	6.5	1.61	456

In the next wells, the drilling process was even more optimized thanks to the wider use of advanced drilling equipment and techniques. For example, rigs with a derrick height of 53 m were used to drill well Nos. 149, 150, and 186, which allowed handling joints with lengths of 35–37 m.

In addition, while drilling these wells, the drillers used equipment such as aluminum DP 147x11 and the ASP-3 automatic pipe handling unit. The rig that drilled well No. 186 was also outfitted with the RPD-type bit feed control unit and ran three-section 9-in. and four-section 7$^1/_2$-in. turbodrills.

The drilling process parameters were monitored and recorded using equipment such as the KPB-3 driller's console outfitted with instruments that indicated parameters of bit feeding, hook load, standpipe pressure, and the RGR-7 induction flow meter and electrical measuring and control instruments. In well No. 186, the interval down to 2148 m was drilled using bits with a diameter of 269 mm. Next a welded liner was run into the well to isolate lost circulation zones. Finally, a 214-mm bit was run and drilled the well to the TD. Bits with a diameter of 269 mm were run mainly in combination with the 3TS5B-9" three-section turbodrills in the bit weight range of 22–37 tons and a mud pump pressure of 120–130 kg/cm^2.

After running in a 400-meter 245-mm liner and isolating the lost circulation zones, the interval below 2148 m was drilled using a 214-mm bit in combination with 4TSSh 7-1/2" four-section turbodrills at a bit weight of 25–30 tons and a pump pressure of 180–190 kg(f)/cm^2. Table 3–7 shows a comparison between the drilling results from this well and well No. 154.

TABLE 3–7
Experimental Well No. 186 Drilled with Modern Drill Rig and ADP Compared with Well No. 154

Well No.	Well depth, m	Number of bit runs	Penetration per bit, m	ROP, m/hr	Bit run speed, m/hr	Overall drilling rate, m/rig • month
154	2,967	169	17.6	8.2	2.45	972
186	2,960	166	17.8	12.8	4.27	1583

This set a record for these types of wells drilled in the Samara region. However, analysis of the drilling process of well No. 186 revealed a potential for further improvement. In the first place, this improvement could be achieved by using effective lost-circulation control methods rather than running in a liner to isolate lost-circulation zones, which would enable running a 214-mm bit. In addition, various auxiliary operations could be done faster. These measures were fully implemented in 1965 when drilling the next well, No. 250. Table 3–8 presents the dynamics of drilling results from all drilled wells.[53]

TABLE 3–8
KTW Drilling Results in Samara Region (Former Kuyibyshevskaya Oblast) at Dmitrovsky Field

Well No.	Drilling method	Bit diameter, mm	Well depth, m	No. of Bit runs	Penetration per bit, m	ROP, m/hr	Bit run speed, m/hr	Overall drilling rate, m/rig • month
			Average results					
For 11 wells	Turbodrill	269	2,959	212	14.0	6.5	1.61	456
			Best results from previously drilled wells					
306	Turbodrill	269	2,955	215	13.6	7.26	1.93	744
344	Turbodrill	269	2,950	215	13.7	6.74	1.90	789
			Results from KTWs					
168	Electrodrill	269	2,970	193	15.3	4.6	1.60	560
157	Rotor	214	2,963	137	21.8	2.6	1.46	590
154	Turbodrill	269	2,967	169	17.6	8.2	2.45	972
149	Turbodrill	269	2,982	199	15.08	11.8	3.2	1,188
150	Turbodrill	269	2,989	205	14.6	11.0	3.2	1,210
186	Turbodrill	269 * 214	2,960	168	17.6	12.87	4.27	1,583
250	Turbodrill	269	3,000	146	20.55	14.85	4.73	2,013

** Two drill bit sizes were used*

Thus, the experience gained from drilling the experimental and KTWs in the Samara region for a three-year period allowed an almost threefold increase in the overall drilling rate in the Dmitrov field. Positive results achieved in well No. 250, attributable to the combination of forced drilling practices and reduced round trip time, were correlated with the results of the optimum drilling parameters calculations for drilling conditions in the Samara region. The researchers who had done these calculations used Equations 3.8 and 3.17 in a mathematical model and accounted for coefficients a, T_0 and 6 for the Podolian, the Vereian, and the Tulian horizons and a rotational speed of 350–450 rpm for the 3TS5B-9 turbodrill. The results obtained are shown in Table 3–9.

TABLE 3–9
KTW Actual Drilling Data Compared with Analytical Studies

Horizon	Interval, m	Well No.	Round–trip time, hrs	WOB, tons		Bit run speed, m/hr	
				Estimated	Actual	Actual	Estimated
Podolskyi	1,355–1,462	250	1.27	40	30	13.0	16.0
		150	1.75	32	18–20	6.4	11.0
Verelskyi	1,579–1,648	250	2.0	38	34	10.8	14.1
		150	2.17	28.7	18	4.7	8.9
Tulskyi	2,138–2,206	250	2.25	42.5	35	3.1	4.75
		150	2.95	29.5	18.20	2.44	4.0

Table 3–9 indicates that the fastest bit run speed was achieved in well No. 250 since the actual bit weight levels in this well most closely matched the estimated levels for this round trip speed. In addition, the estimated bit run speed was quite close to its actual level, which proved the validity of the suggested mathematical model of a well borehole deepening process.

An analysis of the results from the work previously described support the following conclusions:

1. The results of KTW drilling in the Samara region convincingly proved that thorough, integrated utilization of advanced drilling equipment and technologies could lead to significant improvement in drilling results.
2. Use of the turbodrilling method, combined with optimized drilling practices when drilling wells up to 3000 m in the Samara region, achieved the best drilling results.

3. In these conditions, full transfer to rotary drilling was not feasible because it would require a significant upgrade of the drilling equipment such as high strength balanced DC and rotary tables and swivels with improved reliability. Use of this drilling method for drilling wells up to 3000 m, even with the upgraded equipment, would not be economical because it could not achieve high enough results to offset the cost of the equipment upgrade. Selective application of the rotary drilling method for drilling lower well intervals may prove cost efficient.

   The experience gained from the ADP application in well No. 157 was extremely important and proved the possibility of successful operation in rotary drilling with its potential fully used by forced drilling practices, high bit weight (applied to tri-cone and jet nozzle bits), high drilling fluid outflow velocity from bit nozzles, and high differential pressure levels.

4. Currently, the electrodrilling application is feasible only for research purposes. Additional work is required to significantly improve operation of the power supply cable as well as the reliability and durability of electrodrills before these machines have a wide commercial application.

The results from drilling experimental and KTWs were widely used for later drilling in the Dmitrov field as well as in some other fields in the Samara region that had similar geological conditions.

Wells drilled in the Bashkiriya region

Well No. 2000 in the Tuimazin Field. In 1964, well No. 2000 was spud-in on the southeastern wing of the Tuimazin structure to make a comprehensive study of a deep earth crust structure in the Volga-Urals oil and gas province. The study included the composition as well as characteristics of the crystalline basement structure. The entire well interval was drilled using the electrodrilling method exclusively to conduct the required research work in crystalline rock. This type of drilling research work had never been done before either in Russia or anywhere else in the world and was, therefore, of great interest. It is described here because the well borehole deepening process confirmed the functions developed during the research work in sedimentary rocks, the results of which are shown in the previous pages.

Drilling through the crystalline basement was carried out in the interval of 1798–4041.2 m. It was mainly composed of non-homogenous intensively interbedded fractured rocks both acidic and basic, such as gneiss, granite, and diorite. Various direction fractures—from horizontal to vertical—were filled with various ferrous and other compounds. Specialists from the Ufa Oil Research

Institute (UFNII) laboratory determined rock mechanical characteristics using the indentation method. Indentation hardness P_{ω} varied between 88 kg/mm^2 and 530 kg/mm^2, whereas the plasticity factor varied between 1.2 and 2.6 and Young's modulus (1.2–4.4)*10^5 kg/cm^2.

TABLE 3-10
Some Data Recorded at the Experimental Well No. 2000, Tuimazinsky Field, Tatariya

Bottomhole depth, m	Penetration per run, m	Drilling time, hr	ROP m/hr	Drillbit type	RPM	WOB, ton	Flow rate, l/sec	Power at the drill bit, kW	Drillbit torque, kg.m	Power used for rock destruction, kW	Torque on the rock destruction, kg.m	Specific torque, kg.m/ton	Specific work for rock destruction, kg.m/cm^3
2,326	4,30	2,0	1,83	1V-12K	680	25	31	90	129	60	86	3,44	171
2,331	5,1	2,67	2,0	1V-12K	680	25	31	111,5	160	81,5	117	4,68	220
2,346	6,68	3,08	1,67	1V-12K	680	25	31	104,5	149	74,5	106,5	4,26	240
2,353	6,4	3,33	2,0	1V-12K	680	25	31	78	112	48	68	4,6	140
2,936	8,0	2,67	3,0	U-12K	680	18–20	31	100	143	70	100	5,26	126
2,944	7,9	2,73	2,89	U-12K	680	18–20	31	100	143	70	100	5,26	130
2,954	7,8	2,92	2,68	U-12K	680	20	31	112	160	82	117	5,08	164
2,962	10,0	3,17	3,16	U-12K	680	18–20	31	97	139	67	96	5,05	114
2,972	10,3	3,42	3,02	U-12K	680	18–20	31	97	139	67	96	5,05	119
3,101	5,0	2,4	2,08	1V-12K	680	20–22	32	90	129	60	86	4,10	155
3,106	6,3	1,92	3,28	U-12K	680	16–18	32	97	139	67	96	5,64	100
3,127	13,8	4,04	3,42	U-12K	680	20	32	104	149	74	106	5,03	116
3,148	6,1	2,5	2,44	U-12K	680	20	32	113	162	83	119	5,10	183
3,154	9,8	3,77	2.6	U-12K	680	21–22	32	96	137	66	94	4,37	136
3,167	11,5	3,73	3,1	1V-12K	680	19–20	32	92	131	62	88,5	4,55	108
3,195	8,7	3,5	2,5	U-12K	680	19,5	32	95	136	65	93	4,76	140
3,582	17,7	6,23	2,84	U-12K	450	20	19,5	57	124	37	81	4,06	70
3,603	11,9	3,22	3,7	U-12K	680	26	35	100	143	70	100	3,85	102
3,615	16,35	7,2	2,3	U-12K	375	26	35	49	127	32	84	3,2	75
3,636	20	8,92	2,25	U-12K	300	25	19,5	41	121	24	78	3,2	58
3,905	6,3	1,81	3.48	U-12K	530	17	19,5	70	129	47	86	5,0	73
3,930	7,8	3,5	2,21	U-12K	346	25	24	52	118,5	27	75,5	3,0	66
3,942	10,0	4,13	2,42	U-12K	406	27	24	65	156	47	113	4,2	104,5
3,963	4,8	1,19	4,07	U-12K	530	19	24	65	116	40	73	3,85	53
3,968	4,5	1,94	2,31	U-12K	406	17	24	57	137	39	94	5,53	91
3,972	4,4	1,35	3,25	U-12K	466	20	24	61	127	40	84	4,2	66
3,977	6,2	2,83	2,19	U-12K	466	17	24	47	98	26	55	3,24	64
3,985	6,1	1,8	3,38	U-12K	680	21	24	90	129	60	86	4,1	96
3,994	4,0	1,33	3	U-12K	450	27	24	78	170	59	127	4,7	106
4,001	3,2	2,16	1.47	U-12K	346	22	24	44	116	26	73	3,3	95

While drilling the interval of 2156–4036 m, the researchers studied the effect of the drilling process parameters on the results of bit runs. For this purpose, they made 95 test bit runs using special instrumentation and recorders to register parameters of the drilling process, electrodrill input power level, amperage, and voltage (Table 3–10). [54]

Actual power consumed by the drilling bit is shown in Table 3.10. It was calculated using the formula, suggested by F. N. Fomenko: [55]

$$N_{DB} = \left[N_{surf} - \frac{1}{10^3}\left(I^2{}_{A} + I^2{}_{B} + I^2{}_{C}\right) R \right] \eta - N_{FR} - \frac{Pfdn}{19.5 \bullet 10^5} \qquad 3.23$$

where

N_{surf}	is power used on the surface by an electrodrill, including losses in the power cable, *KW*
I_{A}, I_{B}, I_{C}	are phase currents, *A*
R	is power cable resistance, om for cable with a diameter of 50 mm² R=0.38L for cable with a diameter of 35 mm² R=0.5L
L	is well depth, km
η	is the electrodrill motor efficiency factor
N_{FR}	is friction losses at no-load shaft run
P	is bit weight, kg
f	is the conventional coefficient of friction in the trust bearing = 0.003
d	is the shaft diameter, mm
n	is the bit rotational speed, rpm

While drilling the well, the drillers ran the E215/8 and E250/10 electrodrills in combination with 295-mm U-12K and 1V-12K drill bits with K-type cutting structures for hard rocks. Figure 3–28 shows an example of the U-12K bit that was used. The drillers used a frequency control unit to control the rotational speed of the electrodrills. By changing frequency and using electrodrills with a different number of poles, the drillers were able to change the rotational speed within the range of 285–680 rpm. The bit weight range was from 5 tons to 30 tons.

Fig. 3–28 Bit U-12K, worked in the interval of 3972.2 – 3976.9 m

Figures 3–29, 3–30, 3–31, and 3–32 show the types of functions of penetration rate, penetration per bit run, and other parameters of bit weight and rotational speed. The researchers determined the optimum levels of the two parameters mentioned that would allow achieving maximum penetration per bit run. The diagrams indicate that these functions agreed with the well bottomhole deepening model suggested in this chapter. A plateau tendency of the penetration rate curve at higher bit weight could probably be explained by insufficient bottomhole cleaning because drillbits were run without jet nozzles.

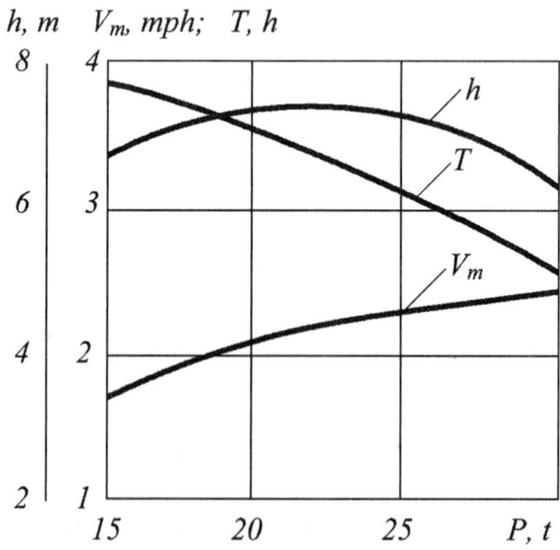

Fig. 3–29 Dependence of penetration (h), penetration rate (v) and bit trip time (T) from weight on bit P at n=680 rpm; Bashkiriya, Tuimazinsky field, well No. 2000; interval 2156–4036 m

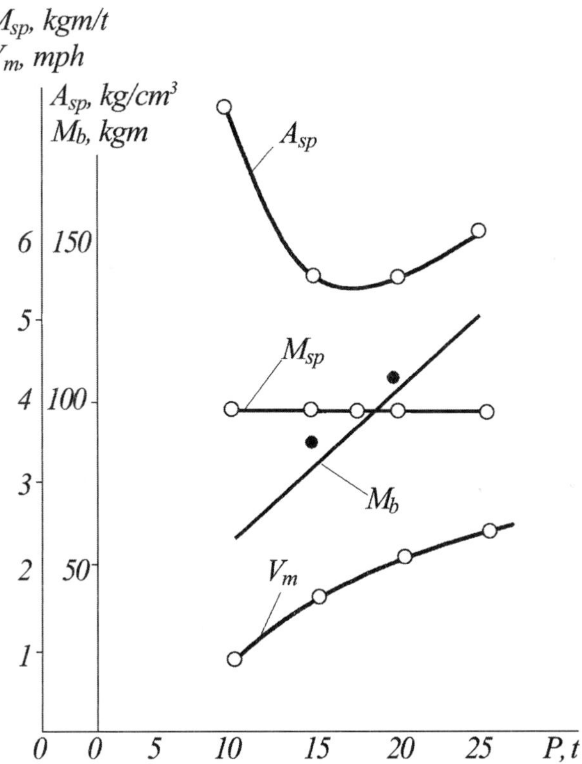

Fig. 3–30 Dependence of specific torque (M_{sp}) and specific power (A_{sp}) for rock destruction, bit torque (M_b) and rate of penetration (V_m) from weight on bit at n=680 rpm; Bashkiriya, Tuimazinsky field, well No. 2000

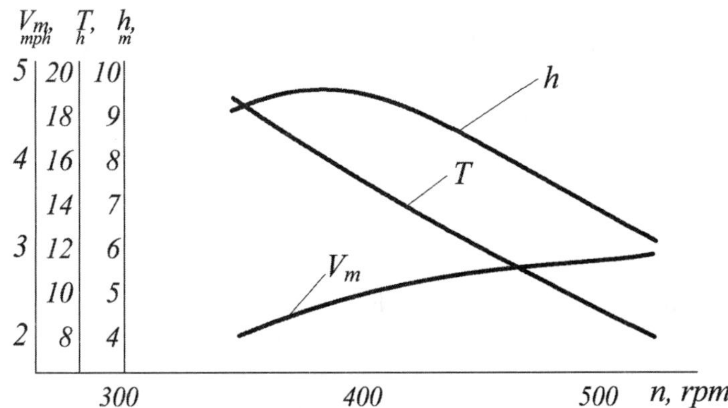

Fig. 3–31 Dependence of penetration rate (V_m), penetration (h) and bit trip time (T) from rotational speed (n, rpm) at weight on bit of 20 tone (interval 3530–3800 m); Bashkiriya, Tuimazinsky field

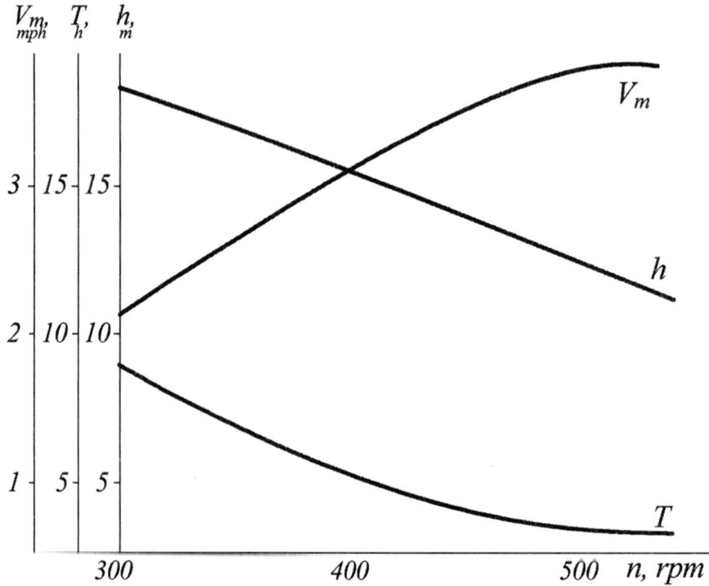

Fig. 3–32 Dependence of penetration rate (V_m), penetration (h) and bit trip time from rotational speed at weight on bit of 25–26 tone (interval 3600–3685 m); Bashkiriya, Tuimazinsky field

At the same time, it should be mentioned that while drilling this well, regular improvement in the drillability of the crystalline rock at greater depths was noted for the first time. This can be explained by the fact that the rock pressure was significantly higher than the drilling mud hydrostatic pressure. This phenomenon was later recognized while drilling the Kola SD-3 well (*see* Chapter 5 in Volume 2).

Small Diameter Wells. [56] In the early 1960s, the drilling industry in the FSU faced the task of increasing the overall drilling speed and reducing the cost of constructing stratigraphic and shallow exploratory wells in remote regions. To achieve this task, specialists from VNIIBT started R & D work in 1962 to improve small diameter well drilling technology, which was especially important for the companies involved in exploratory drilling.

The researchers selected drillbit No. 7 (according to the Russian bit size classification) with a diameter of 161 mm that was commercially produced by the industry at that time. This type of bit featured improved strength compared with bits No. 5 and 6 (140 mm and 125 mm) that were normally used in stratigraphic well drilling. At the same time, bit No. 7 could be used in combination with lightweight DP, small diameter DRC, and low capacity drilling rigs. The Birsk geological-exploratory department (GPK) of the Bashneft Company began test drilling a stratigraphic well with a depth of 1800–2500 m. This type of well was normally drilled using bits No. 6B in combination with the TC4A-5" two- or three-section turbodrills mentioned earlier. Nowadays drillers use PDMs for this type of application.

By utilizing No. 7 bits and a combination of turbodrilling (to a depth of 1300–1500 m) and rotary drilling (to TD) as well as the optimized drilling parameters, drillers were able to achieve significantly higher results in a number of wells. Table 3–11 presents a comparison of these results and the best results from the previously drilled well No. 14 in which No. 7 bits were used.

TABLE 3–11
Small Diameter 161-mm Experimental Prospecting Wells Drilling in Bashkiriya Compared with Existing Well No. 14, Turbodrilling Up to 1300–1500 m Depth, Then Rotary Drilling

Parameter	Well No.					
	30	32*	32	39*	39	14
Well depth, m	1,828	1,833	2,424	1,900	2,283	1,889
Number of bit runs:	120	137	197	207	247	237
including full-size drillbits	103	112	150	169	198	190
Penetration per full-size bit	17.21	14.7	14.4	10.3	10.51	8.87
Penetration rate, m/hr	4.66	4.14	3.37	6.17	5.06	5.68
Bit run speed, m/hr	1.98	1.92	1.38	1.62	1.4	1.64
Overall drilling rate, m/rig month	608	840	538.7	585	457	472

**Additional results for wells No. 32 and No. 39 are shown to a depth of 1833 m and 1900 m respectively for comparison with well No. 14.*

While drilling the test wells, the following equipment and instruments were used.

1) BU-50Br1 drilling rig with a diesel-electrical drive, two B14/200 mud pumps, and all other necessary equipment. The Giproneftemash Design Institute specially developed the rig with a capacity rated at 50 tons. Several rigs of this type were built as a trial series for the stratigraphic well drilling application. The rig consisted of six mobile modules, which greatly facilitated its transportation in areas with a complicated landscape pattern. The rig featured improved operating parameters, which allowed successful drilling of test wells and was equipped with air, electrical, and mechanical control of various units.

2) DRC with a diameter of 127/67 mm and 140/74 mm.
3) External upset DP from steel grade D with a wall thickness of 8 mm butt-welded to ZU-120 tool-joint with a high performance premium thread.

The rigs primarily used water as a drilling fluid. When running turbodrills, the drilling parameters were WOB 8–10 tons with a flow rate of 15–18 liters/sec. For rotary drilling, the parameters were WOB 8–14 tons, flow rate 16–19 liters/sec, and rotational speed of 60–90 rpm. The experience gained from drilling these wells was greatly extended in the Birsk GPK.

While drilling well Nos. 39 and 30 using the rotary method, the researchers carried out work to determine the dependence of penetration rate from bit weight level, other parameters being equal, using Lubinsky's method previously mentioned.

Drilling wells in the Urals-Volga oil and gas province using dynamic processes to intensify rock destruction at the bottomhole[57]

The theoretical and experimental work carried out from 1970 through 1979 by research engineers from the VNIIBT institute revealed that the intensity of the mechanical destruction of rock using cone bits depended greatly on the dynamic load level applied to the rock.

The researchers conducted the test work in the lab using special test bench machines. In addition, full-scale tests were performed in a special well simulator of the SKTBE in Kharkov, the Ukraine, while running the E-185 electrodrill machine with a rotational speed of 220 rpm and 680 rpm combined with the 2K214SG and 2K214TKZ cone bits. [58]

The STE-1I downhole telemetry system provided information about the drilling process. The DS included DP with a diameter of 140 mm with a power cable inside. Remote sensors measured axial forces $P(t)$ in the DS cross-section above the electrodrill during continuous increase and decrease of a bit weight. This allowed establishing a dynamic load P dependence from the variation of static load P and determining the amplitude and range of the dynamic load.

Russian specialists carried out a certain amount of research work and reviewed a number of studies related to the phenomenon of non-linear propagation of oscillations in rods with a gap in a particular area of the rod connection. As a result, they concluded that this phenomenon, increasing mechanical resistance of oscillations propagation in the string (the so-called impedance), allowed them to increase the bit weight dynamic component and achieve a significantly higher penetration rate and penetration per bit run.

The studies [59] [60] present descriptions of the DS divider designs developed by engineers from the VNIIBT, along with their test results and theoretical studies of the divider optimum position in a DS. The tests were carried out in a SKTBE simulator well in Kharkov, the Ukraine, as well as in a commercially drilled well in the Abdrakhmanov field in the Tatariya region. The tests proved their conclusion about bit performance improvement.

At the same time, the tests revealed that the various designs of DS dividers could not be used widely by the industry because of their complex design and low durability. Then the idea came to use DP made of different materials to achieve a non-linear effect, rather than using a divider to form a gap section. For this purpose, a number of steel DP were installed above the DHM, and the remaining part of the DS was made up of ADP. According to information from the study, [61] the distance between the transitional section and the bit should be equal to the odd number of parts representing a quarter wave length of the bit oscillations frequency component, i.e., this distance should provide the maximum DS impedance level. When using this DS design, engineers have to observe the parameters of the drilling process, used for that distance calculations.

The main volume of research works, testing, and implementing the technique were arranged in Tatneft oil association. They ran the 3TSSh-195 turbodrills in combination with 214-mm bits that were normally used in that region. The research engineers developed the final DS design faced some problems related to the utilization of such DS commercially. One of the problems was the necessity to maintain the required static load level and the proper level of the entire drilling process because there was only one boundary between the steel and aluminum materials. The final design included the broadband reflector, which in a certain DS section consisted of a set of alternating steel and aluminum DP.

The experience in drilling the Tatneft wells using a controlled dynamic bit weight indicated that the average penetration per bit increase was 15% in 1975 at an annual drilling volume of 500,000 m. Further improvement of the method may allow an additional increase of this parameter by 30–40%. According to the data presented in the work, [62] this method was used by drillers from the Tatneft Company since 1972 and saved 2000 bits. Also, the same year, drillers from the Permneft Company began using this method.

The theoretical and test research work showed that assessment of DS dynamics allowed control of a drilling process by registering elastic oscillations formed as a result of bit operation and propagation in the DS and the rock. This assessment may also obtain data about bit rotational speed, physical and mechanical characteristics of drilled rock, expected problems, etc.

Drillers sometimes use information on elastic oscillations propagation obtained from monitoring bit performance to determine coordinates of a bottomhole position without running special measuring instruments into a well. [63]

When using the rotary drilling method, drillers in the United States control the dynamic load applied to a drillbit. As for drilling using the DHMs, a control similar to this has never been studied in the West. Considering the fact that DRCs are used widely by Western drillers with diameters that are significantly different from those of DP and that the transition boundary also serves as a wave oscillation reflector, this type of DS may be regarded as a means to control the dynamic load applied to a drillbit. Calculations indicated that the ratio between DRC length and bit rotational speed used in U.S. drilling practices stayed within a range that ensured the maximum dynamic load level. This was due to the maximum impedance of a DS bottom section. Although there are no specific indications in related studies and reports that show a conscious approach of the Western drilling industry to select the DS bottom section design in order to achieve maximum mechanical resistance of the DS bottom section, the coincidence should not be regarded as accidental.

The results of the theoretical and experimental studies related to use of elastic oscillations generated in a DS, surrounding media, and rocks indicated that this development trend held promise for improvement of drilling process efficiency.

According to available information, American and French researchers started obtaining data about downhole dynamic processes using acoustic measurements taken at the surface.

Experience from KTW Drilling and Results of Its Implementation

After construction of the test wells was completed in Azerbaijan, the Samara region, and the Bashkiriya region, the suggested mathematical model of the well borehole deepening process was tested and experience was accumulated. Researchers, in conjunction with drillers, started the next phase of this work that was related to company-wide use of KTW drilling results for development of the entire field or region Rather than for individual wells. This method was especially efficient when used for drilling KTWs immediately after the first exploratory wells that proved the presence of oil-bearing formations in a field at the drilling delineation well stage. In this case, the results and experience from the KTW construction could be used for drilling all other wells in the field, which, unlike with the existing practice, immediately increased the average drilling penetration rate in the field.

In this respect, a good example is the experience and results from drilling KTWs in the Kudinov field of the Volgograd region where the drilling was carried out by Archedin UBR (drilling directorate) of Nizhnevolzhskneft association. During a period of several years, the primary results from wells drilled by this company improved twofold, which is shown in Table 3–12.

TABLE 3–12
KTW Drilling Results in Nizhnevolzhskneft at Kudinov Field

	1965		1966		1967		1968		1969	
Results	Average from drilled wells	From KTW No. 90	Average from drilled wells	From KTW No. 108	Average from drilled wells	From KTW No. 98	Average from drilled wells	From KTW No. 15	Average from drilled wells	From KTW No. 8
Depth, m	3,360	3,210	3,247	3,170	3,243	3,458	3,189	3,193	3,225	3,200
Total footage drilled, m	45,426	–	57,518	–	63,258	–	66,396	–	67,923	–
Number of drilled wells	13	–	18	–	20	–	21	–	21	–
Penetration per bit, m	9.14	14.27	11.51	14.95	13.61	19.32	15.98	20.0	17.04	19.4
Overall drilling rate, m/rig•month	360	456	493	688	526	720	575	912	670	948
Average annual drilled footage per drilling crew, m	4,130	–	4,793	–	5,751	–	6,640	–	8,490	–
Cost per meter, rubles	117.46	115.8	108.83	101.2	100.5	83.5	94.0	79.09	93.7	78.25

The table presents the best results from wells drilled each year. Actually, each year drillers constructed two to three KTWs. Also, it should be mentioned that significant improvement of drilling results was achieved as early as 1966 when the registered average penetration rate exceeded the penetration rate in well No. 90, which was drilled in 1965.

Continuous efforts of specialists to implement the most positive experience from drilling KTWs for the entire field development achieved annual improvements in drilling results. In 1969, the efforts of the specialists from the VNIIBT and VolgogradNIPI research institutes jointly with the engineers from the UBR and association showed an annual penetration increase of 1.5 times for the UBR compared with the results from 1965. During the same period, the number of drilling crews was reduced from 11 to 8 while average annual penetration per crew increased more than twofold.

Utilization of the conventional trial-and-error method of drilling technology improvement would, in the best-case scenario, result in improvement of drilling results by 30–40% (which does not happen very often). However, the Archedin UBR Company in the Volgograd region used the experience from KTW drilling in 1969 and saved nearly 1,000,000 rubles. This economic effect was achieved by using drillbit types selected for the corresponding lithological characteristics of rock in the entire well section, running three-section turbodrills instead of two-section and optimizing the drilling process parameters. Additional benefits were achieved by using aerated drilling mud while drilling a large part of the well interval and other innovative techniques that had been tested earlier in some KTWs.

Similar work was carried out in the Antipov-Balakleyev field by specialists from the Nizhnevolzhskneft association, in the Rechitsa field by the Belorusneft association, and in the Orenburg and Shatlyk gas fields, as well as in other fields.

The nominal drilling speed levels achieved in the Kudinov field, for example, may not seem as high when compared with the results from drilling wells of this depth in other oil producing countries around the world, such as the United States. However, it should be noted that the geological section in the fields of the Volga-Urals province is mainly composed of hard and very hard rock of the Paleozoic sedimentary cover. Yet, the most important negative factor, as mentioned in Chapter 1, was poor management of the drilling operations. This led to significant time losses due to continuous shortage of required equipment and materials, poor condition of access roads, low durability of certain pieces of equipment (especially mud pumps), as well as bad planning, and poor motivation for drilling crews. By eliminating all these problems, drillers could easily improve the results by 1.5–2 times.

These problems did not affect the importance of adopting the method of drilling process optimization using experience from KTW drilling, which, undoubtedly, would be as efficient when used under other conditions of drilling process management.

In 1969, scientific and research institutes in six regions of the FSU carried out this work, using as a guide the first edition of *Technique for KTW Drilling,* published in 1968.

In 1970, the Board of the Ministry of Oil Industry reviewed the joint report by the VNIIBT and Nizhnevolzhskneft Company on results of KTW drilling in the Kudinov and the Antipov-Balakleyev fields of the Volgograd region. The Ministry's Board approved this experience and issued the Order of the Ministry of Oil Industry No. 384, dated July 29, 1970. The Order instructed all drilling companies, in cooperation with the regional scientific and research institutes, to carry out work for the development and commercial use of well drilling technology using the experience from KTW drilling.

New drilling technology and equipment used for drilling KTWs achieved significant cost and time savings; even though the planned R & D work to be done by the specialists from VNIIBT required additional time.

A typical example of such work is drilling KTWs in the Orenburg gas field. In 1971, the average overall drilling rate was 454 m/rig-month, whereas the maximum level of this parameter, achieved in well No. 175, was 465 m/rig-month. In 1971 and 1972, drillers achieved the following levels of overall drilling speed while drilling several wells:

No. 170 655 m/rig-month;

No. 174 675 m/rig-month;

No. 177 886 m/rig-month;

No. 166 1028 m/rig-month.

During the same period, penetration per bit increased from 26.9 to 39.2 m, and the penetration rate increased from 3.55 m/hr to 7.75 m/hr.

While drilling the interval from the surface down to 3000 m in KTW No. 3 in the Severnaya field operated by the Krasnodarneftegaz Company in 1974, drillers achieved a reduction of the overall drilling time of 41 days. This compared to 89 and 73 days respectively to complete the same interval in well No. 1 and well No. 2 in the same field. Faster drilling speed in the interval from the surface down to 3107 m made it possible to set one string without running a conductor casing down to 1100 m. In well No. 3, drillers reached 5000 m in 9 months compared to 14.5 months for well No. 1. In addition, the number of bits per well went down from 188 to 130.

The number of drillbits used for drilling KTW No. 22 and other wells in the Krasnoborsk field operated by the Belorusneft Company was reduced to 37 from 67, whereas the overall drilling rate increased from 532 m/rig-month to 1017m/rig-month. In 1974 the economic effect of implementing the new technology totaled 1,037,000 rubles.

By January 1, 1976, a total of 132 KTWs had been drilled in the fields operated by 14 companies, including 93 wells drilled from 1971 through 1975. During this five-year period, drillers used the optimized well drilling process based on KTW drilling to drill 2,000,000 m. According to incomplete data from annual reports by the NIPI research institutes, the total economic effect for the increase was 16,700,000 rubles.

In 1976, VNIIBT published a third edition of *Technique for KTW Drilling*, approved by the Ministry of Oil Industry, which covered all main well drilling operations. By that time, a number of regional NIPI institutes in cooperation with drilling companies on their own initiative extended the scope of research work carried out while drilling KTW. This was not limited to the issues of selecting drillbits, drilling methods, and drilling practices. For example, while drilling KTWs in the Kudinov field, drillers started using aerated drill mud and humate-potassium instead of coal-alkali for drilling fluid. The well designs were changed in a number of regions.

This advanced experience was summarized and developed in a new technique, which was introduced in 1976 and 1977. Table 3–13 indicates that for a period of four years from 1976 through 1979, companies operating under the supervision of the Ministry of Oil Industry drilled 124 KTW. According to information from the NIPI institutes, recommendations that were based on the analysis of results from drilling KTW were implemented, and drilling wells with a total of 1.5 million meters achieved an economic effect in excess of 20.0 million rubles.

TABLE 3–13
Number of KTW Drilled in 1976–1979

Oil company	Area of activity	Research organization	1976	1977	1978	1979	Total
Nizhnevolzhskneft	South Volga, North Caspian, North-West Kazakhstan	VolgogradNIPI	3	2	2	4	11
Komineft	North-East Russia	PechorNIPIneft	6	2	2	2	12
Stavropolneftegaz	North Caucuses	SevKavNIPI	2	2	2	1	7
Ukrneft and Belorusneft	Ukraine and Byelorussia	UkrGiproNIIneft	3	3	4	2	12
Permneft	Volga-Urals – North	PermNIPI and Perm Division of VNIIBT	7	3	2	3	15
Glavtyumenneftegaz	Western Siberia	SibNIPI	7	2	4	5	18
Bashneft	Bashkiriya	BashNIPIneft	2	3	4	3	12
Kuibyshevneft and Orenburgneft	Volga-Urals – South	GiproVostokneft	3	3	4	4	15
Saratovneftegaz	Middle-Volga	SKB Saratovneftegaz	1	2	3	3	9
Sakhalinneft	Sakhalin island		2	3	No data available (NDA)		5
Krasnodarneft	South Russia	TermNIPIneft	NDA	1	2	1	4
Turkmenneft	Turkmeniya	TurkmenNIPI		2	2	NDA	4
All FSU							124

For the period from 1968 (publication year of the first edition of *Technique for KTW Drilling*) through 1980, the number of KTW was calculated to be around 300, and the new techniques based on KTW drilling results analysis were implemented in wells with an overall penetration of about 4,000,000 m. This had an economic effect of nearly 40.0 million rubles.

At the same time, analyses indicates that KTW efficiency can be substantially improved by increasing amount of research work in such trends as well completion, well designs, elimination of downhole problems, etc., which have not been addressed sufficiently by NIPI research institutes jointly with the operating and drilling companies.

In 1981, the Ministry of Oil Industry published the fourth edition of *Technique for KTW Drilling*. The book was designated as a Management Directive of industrial level RD-39-2-642-81. More than 20 publications related to the issue of KTW drilling were released from 1971 through 1980 and confirmed that a large volume of related research and experimental work was being carried out jointly by scientific and research institutes and drilling companies.

The information presented in this section offers the following conclusions:

1. More than 10 years of experience acquired by drilling KTWs in various regions of the country, along with the results from implementing recommendations based on the experience, convincingly proved the feasibility and high efficiency of this technique of development well drilling optimization. It allowed faster development of fields containing hydrocarbon reserves.
2. Researchers who carry out experimental work and studies while drilling KTW should focus on obtaining information related to well design, drilling through productive zones, well completion, and elimination of downhole problems to streamline the entire process of drilling these wells.
3. Drillers should continue to use positive experience from small diameter running drillbits of 161 mm, and a combination of rotary and DHM drilling methods as well as develop and use the experience for exploratory drilling in regions using mobile drilling rigs of modern design.

To summarize this section, this trend of R & D work was continuously slowed down and finally terminated in the 1980s. This occurred due to a series of changes in the organizational structure of the Ministry of Oil Industry and of scientific and research organizations. The changes included reshuffling of management and dissolution of the Department of Drilling Technologies of the VNIIBT. The latter ceased performing its managerial and coordinating functions related to regional scientific and research institutes in KTW drilling, which is a perfect example of the negative tendencies that were mentioned in Chapter 1.

Areas of Feasible Application of Various Drilling Methods—Rotary, HDHM, and Electrodrilling

Drillbit rotational speed as optimization criteria

For a long period of time, the FSU remained the only country in the world where drillers used these three methods described for commercial well drilling. During the same period, various opinions on the issue were expressed and advocated.

At the same time, a proper scientific approach to this issue and to the entire process of well drilling based on the analysis of KTW drilling results Rather than on arbitrary decisions of a few individuals may allow the elimination of the uncertainty. A vast amount of experience accumulated during KTW drilling using the three drilling methods made it possible to perform the required analysis and work out specific recommendations related to optimum application for each of the three methods.

The main criterion for selecting a drilling method was the optimum bit rotational speed that, along with other optimum parameters, enabled achieving maximum bit run speed or minimum cost per meter.

The optimum drilling parameters were dependent on the bit type, rock physical and mechanical characteristics, drilling mud properties, and well depth. This last factor determined geological conditions of a section as well as the time required for round trips and auxiliary operations. A certain optimum proportion of these drilling parameters can be achieved. The selected drilling method, as well as drilling equipment and tools for its implementation, must meet these requirements.

Accordingly, researchers developed this procedure. Optimum drilling parameters and downhole power characteristics were determined while drilling KTW. Next, they were used to select the optimum drilling method, and finally they were used to determine the DP and DRC parameters, drilling rig, and all related surface equipment.

The experience gained from drilling KTW and analysis of the well deepening regularities showed that drilling parameters should change as the well depth increased. We have taken into consideration that when the well depth increased, more time would be required for round trips and auxiliary operations, and its proportion to bit on bottom time would increase consequently, that is why the drillbit rotational speed should be lower to achieve greater bit-on-bottom time and finally better bit run speed.

$$V_r = \frac{V_m}{1+\varphi} \qquad 3.24$$

where

V_m drilling penetration rate, m/hr

φ a non-dimensional parameter, which is a proportion between cumulative time for round trips and related auxiliary operations T_{trip} and bit-on-bottom time T_{dr}

$$\varphi = \frac{T_{trip}}{T_{dr}} \qquad 3.25$$

Figure 3–33 indicates that at small well depth levels, when φ is small ($\varphi \leq 3$), the penetration rate factor plays a major role for V_r increase, since, in this case, V_r strongly depends on V_m. As the well depth and φ increase, further growth of the penetration rate would no longer be a decisive factor. To achieve φ reduction, drillers should increase penetration per bit and bit-on-bottom time. Accordingly, drillers should run high-speed DHMs in the upper intervals, slower DHMs in the middle, and use rotary drilling to drill deep intervals.

When PDC and diamond bits are used, the φ parameter is quite low and penetration rate becomes the main factor determining the efficiency of running these bits, which prompted the use of diamond bits in combination with high-speed DHMs. We found this effect of well depth on variation of the optimal bit rotational speed during the analysis and mathematical processing of results from drilling large numbers of regular wells using rotary and electrodrilling methods in the Kotur-Tepe field operated by the Turkmenneft Company in 1969 and 1970. [64] The analyzed data were collected in the interval 2200–3700 m, composed of red beds, at bit rotational speed changes from 70 rpm to 680 rpm for the B11S type bit, and with constant bit weight in the range of 14–17 tons. Figure 3–34 shows the results of processing data from 500 bit runs.

The diagrams in Figure 3–34 indicate that the optimum levels of a tri-cone bit rotational speed, required to achieve the maximum penetration per bit, stayed within the range of 70–80 rpm for any well depth. Whereas, to achieve the maximum bit run speed, they changed from a maximum level of 325–350 rpm at a depth of 2000 m to lower levels at greater well depths. At 4000 m, the optimum bit rotational speed was 250–260 rpm and decreased to 80–100 rpm at greater depths.

The accumulated experience allows a conclusion that, considering the great variety of drilling conditions in Russian oilfields and the entire world, the optimum

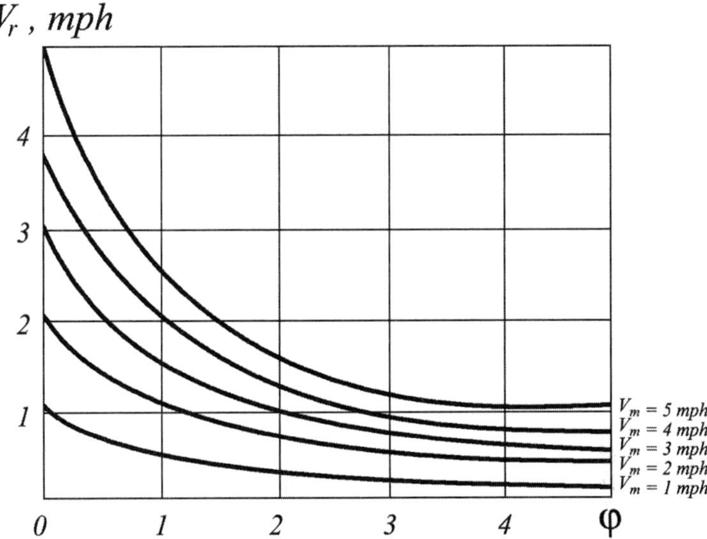

Fig. 3–33 Dependence of bit run speed (V_r) on parameter φ (well depth characterization factor) at various penetration rates (V_m)

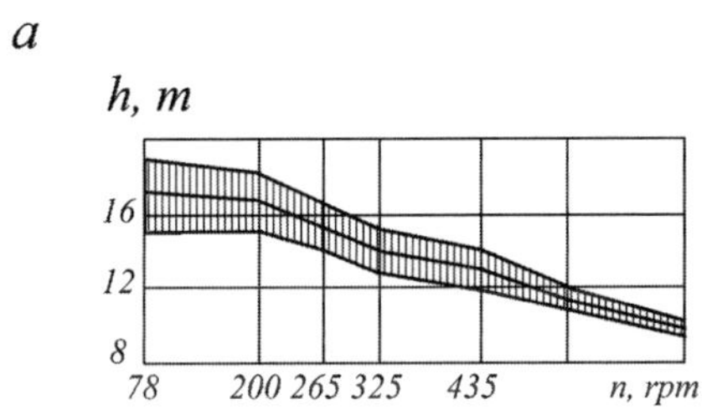

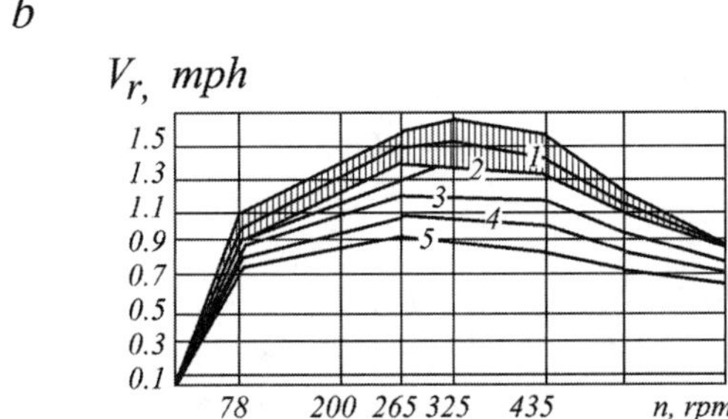

Fig. 3–34 Dependence of drilling performance from rotational speed at various depths:
a) penetration b) bit run speed
1, 2, 3, 4, 5 – 2000, 2500, 3000, 3500, and 4000 m correspondingly
Kotur-Tepe field, Turkmeniya

level of bit rotational speed varies from 40–50 rpm to 700–800 rpm. Accordingly, use of the following drilling methods and DHMs is recommended, depending on the optimum levels of bit rotational speed (rpm):

Drillbit rotational speed, rpm	Rotary drilling	Drilling with DHM
40–100	Yes without limitations	Electrodrill or turbodrill with two-step gear reducer
100–250	Yes limited with the well depth and equipment available	PDM, electrodrill, or turbodrill with gear reducer
250–500	No	Three-section spindle-type turbodrill with a precision-cast turbine, or turbine with a pressure line falling towards the break; electrodrill with a gear reducer; super high-speed turbodrill with gear reducer
500–800 and more	No	Turbodrills and electrodrills for diamond drilling application

Hence, when designing an optimum drilling process, a rotary drive and a DHM should be regarded as tools that complement each other rather than fully opposite tools with significantly different levels of rotational speed, although this was the case at the initial stage of running turbodrills. Drillers now have a new paradigm of optimization as a possibility for selecting a drilling method and the corresponding equipment using the optimum drilling parameters for given conditions, rather than vice versa.

Test wells drilling in 1979–1980

The principle of gradual reduction of a cone bit rotational speed with depth and the proper combination of various types of DHMs with the rotary drilling method is also true for utilization of modern design low-speed cone bits with sealed journal bearings.

In 1979 and 1980, this principle was proved by drilling special experimental wells in the fields operated by Samaraneft in the Volga-Urals and Surgutneftegaz in the Western Siberia oil provinces using a combination of rotary drilling with various types of HDHMs. [65] This work was also necessary in light of the fact that introduction of low-speed bits in the FSU triggered a "rotary drilling mania," a new campaign for using the rotary drilling method exclusively. During this campaign, the Ministry of Oil Industry decided to completely transfer to the low-speed drilling technique in all regions of the country including Western Siberia. To do this, imported high strength DP and DRC and modern drillbits were supplied to regions where drillers normally used turbodrills, such as the Samara and Orenburg regions, the Surgut area of the Tyumen region, etc. At that time, this equipment was limited in supply, and drillers needed them to drill deep wells in other regions.

In addition, the Ministry significantly cut the supply of ADP, turbodrills, and spares for them to the regions where the turbodrilling method was predominantly used.

Table 3–14 and Figure 3–35 show the results of drilling the experimental wells in Samara region at Tverskaya field: experimental well No. 163 using the turbodrilling method and well No. 176 using the rotary drilling method, as well as Karagaisk field well Nos. 178, using rotary drilling with imported bits.

TABLE 3–14
Experimental Wells Drilled in Samara area by Kuibyshevneft Company in 1979–1980 (see Figure 3–35)

		Tverskaya field		Karagaisk field
Parameters	Units	Well No. 163, turbine drilling	Well No. 176 rotary drilling	Well No. 178, rotary drilling with imported tri-cone bits
TVD	m	3,164	3,104	2,855
Number of drillbits		76	33	14
Drillbit run	m	41.6	94.1	267.1 (8½" bit)
ROP	m/hr	10.97	4.74	6.85
Drillbit run speed	m/hr	4.10	3.35	5.41
Overall drilling rate	m/rig-month	1,910	1,658	1,982
Total time on well, including:	hours	1,337	1,345	1,037
Machine drilling	hours	308	655	416
Tripping	hours	464	271	109
Casing	hours	110	47	134
Preparation/auxiliary	hours	436	339	241
Down-time	hours	19	33	137

Table 3–15 presents data from Fedorovsky field developed by Surgutneftegaz: well No. 1825 with vertical borehole drilled using the rotary method in compare with results from directional wells Nos. 2873 and 2875 drilled using the 3TSSh1-195TL turbodrills.

TABLE 3–15
Experimental Wells Drilled on Fedorovsky Field in Western Siberia by Surgutneftegaz Company in 1979–1980

		Cluster 131		Cluster 107
Parameters	Units	Well No. 1825, rotary, vertical	Well No. 2873 turbine, directional	Well No. 2875, turbine, directional
TVD	m	2,384	2,400	2,566
Number of drillbits		5	9	10
(one is 11⅝" and other 8½" bits)		(11⅝" bit run with turbine)		
Average drillbit run	m	476.8	266.7	256.0
Average 8½" bit run	m	495.8	249.3	240.2
Average ROP	m/hr	21.82	35.29	35.64
Average 8½" bit ROP	m/hr	18.96	31.24	32.65
Drillbit run speed	m/hr	13.66	19.50	19.00
Overall drilling rate	m/rig-month	4,918	7,059	7,128
Total time on well, including:	hours	349	244	260
Machine drilling	hours	109.3	68.0	72.0
Tripping	hours	27.0	28.0	35.0
DP making-up	hours	39.0	27.0	28.0
Casing	hours	74.6	55.0	46.0
Preparation/auxiliary	hours	62.2	56.0	75.0
Down-time	hours	36.9	10.0	4.0

Rigs that drilled the experimental wells were outfitted with all modern equipment and used advanced technologies, which allowed maximum use of each method potential and showed better drilling results. This is indicated by the results that were achieved by drillers from the Samaraneft Company and compared with the best results from wells drilled previously. This proved that the drillers selected the right equipment and technology for those particular applications.

Drilling the experimental wells showed that DHMs used with Russian bits still allowed drillers to achieve faster bit run speed and lower cost per well drilled, even though fewer bit runs were made when the rotary drilling method was used in combination with modern design domestic drillbits or even more durable

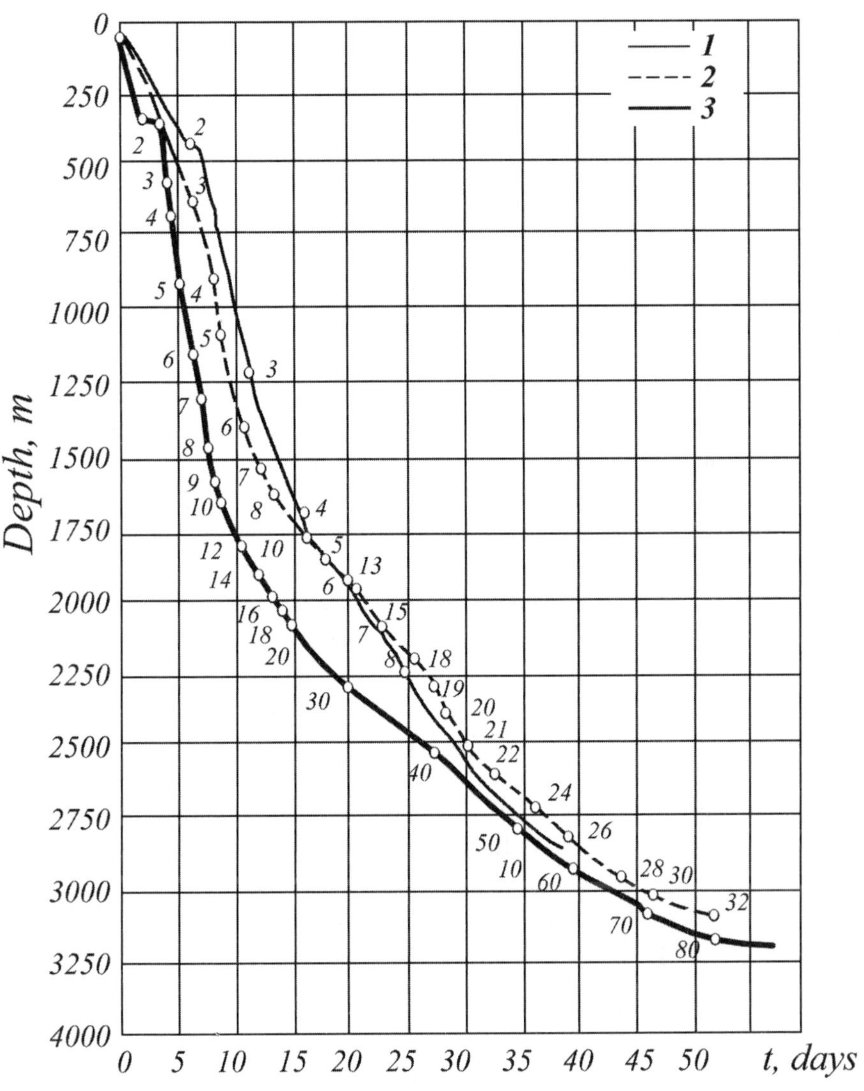

Fig. 3–35 Drilling times of wells No. 163, 176, and 178, Samaraneft Company:
1 – well No. 163, Tverskaya field (turbodrilling method)
2 – well No. 176, Tverskaya field (rotary drilling method)
3 – well No. 178, Karagaisk field (rotary drilling method, Western bit)
4 – well No. 178, Karagaisk field (rotary drilling method, western bit)
(Numbers at the points on graphs represent how many bit runs)

foreign-made bits. This fact supported the conclusion shown regarding the effect of a well depth factor and related non-dimensional parameter φ. At $\varphi \leq 3.0$, the effect of the penetration rate predominates, and drillers should use a drilling method that achieves the highest level of this parameter. At higher φ, drillers should choose a method that enables the achievement of longer bit-on-bottom time and greater penetration per bit run. Hence, in most cases DHM application is feasible for drilling wells with depths of 3000–3500 m, whereas the rotary drilling method or DHMs with gear reducer, and multi-sectional turbodrills should be run in deeper wells.

One more important conclusion follows: the introduction of low-speed bits in no case means a rejection of standard Russian bits use with an unsealed bearing. These bits are still required for drilling upper well intervals in combination with high rotational speed motors (n=350–400 rpm) because they can achieve high penetration rates and well deepening speeds at minimum cost due to their low price.

A growing number of complaints from drilling companies and toolpushers working in Western Siberia in particular, forced the abandonment of this old-fashioned practice. At the same time, a positive outcome of the commercial production of tri-cone bits with a sealed bearing at the Samara bit plant achieved very good results by selectively using these bits in applications where they were best suited.

Some positive examples of tri-cone bit selection tendencies were demonstrated later in the mid 1990s (*see* Chapter 2) when the Russian market opened to different techniques and operators were able to optimize drilling practices in the new economic conditions. They still use high-speed domestic made tri-cone bits along with foreign made low-speed ones. Furthermore, all major bit manufacturers were offering high-speed (200–220 rpm) journal bearing sealed tri-cone bits in the late 1990s to optimize performance, which fully corresponded to the study made more than 20 years earlier.

In this section, we have analyzed only one criterion for a drilling method selection—bit rotational speed. A number of other factors, such as a well profile (vertical or directional and horizontal), also affect selection of one drilling method or another. DHMs, especially those types that could be run in combination with a telemetering system, such as electrodrills, are most efficient in directional and horizontal well drilling applications. A rotary drive or a turbodrill with gear reducers show best results in deep well intervals due to higher temperatures. Turbodrills cannot be used when weighted drilling mud with a density exceeding 1.7–1.8 g/cm^3 is required or if air must be used to clean drill cuttings from the well borehole.

Drilling research specialists, using the experience from drilling KTW and various experimental wells and the analysis of all factors affecting the selection of a drilling method, determined the following areas of efficient use of the three drilling methods in question. [66]

Typical areas of application for rotary drilling

Select the rotary drilling method for the following conditions:

- running cone bits in deep well intervals with depths greater than 3000–3500 m when bit-on-bottom time is a significant factor and the bit rotational speed does not exceed 100 rpm
- drilling thick layers of plastic clays, dense shale, and other type of rock in which blade-type or tri-cone bits with long milled teeth show best results in combination with high velocity fluid outflow from jet nozzles (100–120 m/sec)
- drilling hard, highly abrasive rock
- running cone bits with a diameter of less than 190 mm (except for multilateral and horizontal wells)
- drilling wells with weighted drill mud with a density of more than 1.7–1.8 g/cm^3 if the electrodrill cannot be used
- drilling wells with high bottomhole temperatures (more than 140–150° C) and serious downhole problems, such as rock sloughing and caving, and major loss of circulation
- regular coring, primarily spot coring
- air and foam drilling when running an electrodrill is not possible
- KTW drilling

Typical areas of application for HDHM

HDHMs show best results in the following applications:

- drilling wells, both vertical and directional up to 3000–3500 m and in some cases even deeper, using 190–445 mm diameter tri-cone bits with drilling mud density up to 1.7–1.8 g/cm^3 (possibly higher with PDM) when optimum levels of bit rotational speed exceed 100 rpm

- drilling with natural diamond and ISM (also PDC nowadays) drillbits in deep wells with mud density of 1.7–1.8 g/cm^3 (with the exception of PDM) and at temperatures up to 140–150° C (if DHM is equipped with rubber coated components)
- inclination angle buildup in directional wells until the well path is stabilized, irrespective of optimum rotational speed levels
- tailing in by horizontal and multilateral wells, as well as sidetracking from existing boreholes, to stabilize and increase production from wells with low flow rates
- upper interval of large diameter deep wells using the RTB double-turbine drills
- foam drilling using low aerated mud to eliminate lost circulation regardless of the rotational speed optimum levels and to improve results of bit runs with optimum rotational speed levels
- cement drilling out in the casing using tubing as a DS

Typical areas of application for electrodrilling

Prior to listing feasible electrodrilling applications, it should be mentioned that the industry saw significant improvement of the design and operational reliability of electrodrills and all auxiliary equipment during the past 15 years since the first experimental wells were drilled with electrodrills in the fields operated by Azneft, Samaraneft, and Bashneft, (for details *see* Chapter 2).

Electrodrills are recommended for the following applications:

- drilling deep wells using weighted drill mud with a density up to 2.3 gr/cm^3 and bits with a diameter of 190 mm and larger when it is necessary to achieve optimum bit rotational speed levels for the entire well interval at bottomhole temperatures of 100–120° C

- drilling directional and vertical wells in combination with telemetering systems, especially wells with complex geological conditions, providing optimal bit rotational speeds for all intervals of the entire well
- KTWs
- drilling horizontal and multilateral wells, especially at depths greater than 2000–2500 m when complicated conditions do not allow running DHM because of difficulties in a telemetering system application
- air, foam, and aerated drilling mud with a high degree of mud aeration
- drilling with ISM and natural diamond bits

Drilling Optimization System Conclusions

Information in this chapter leads to the following conclusions:

1. The availability of three drilling methods is a great advantage for drilling companies and promotes further development of drilling equipment and technologies.

   The objective is to find the proper application for each method, use its maximum potential, and make every effort to develop it.
2. Development of the well construction optimization process using KTW techniques is probably the fastest and most effective way to achieve the best possible conditions for development of a field in terms of time and money.
3. Development of a field by a single contractor creates the most favorable environment to realize the potential of the KTW method. If this is not possible, then it is suggested that the continuity of experience and results of KTW drilling be made available to all contractors.
4. The lightweight mud application, including water, provides the best results in combination with high-speed DHM. This has to be accounted for by the fact that the ROP has a linear dependency on bit rotational speed when differential pressure decreases.

5. The ROP of tri-cone bits can be significantly increased in hard rock by the installation of the proper dynamic wave reflectors. These are the waves that are created by the interaction of the bit with the rock.
6. In principal, use of bits with sealed bearings, which allow increasing efficiency of low rotational speed drilling, does not change the conclusions about the usefulness and spheres of efficient application of the three drilling methods in question.

To summarize, the conclusions we should evaluate are whether the KTW technique was just a historical example or whether it could be useful in the frame of modern trends in the drilling industry. Consider the statement by Keith Millheim in the *Oil & Gas Journal*, September 17, 2001: “Proactively managed drilling operations optimize company performance.” [67] The example offered by the author of a proactive company was an aggressive approach in a new drilling situation for high-risk wells when significant efforts in data acquisition and analysis were made with the first two wells. Further, this achieved very sound time savings when the next four wells were drilled. This reflected the KTW idea very well. What was called a “drilling performance system” (*see* Figure 3 in the *OGJ* article) actually could be considered as KTW techniques with up-to-date equipment and information technologies, which would provide much faster, and finally, on-line analysis and decision making.

We also believe that the information collected during KTW drilling would be of help for the operators in the areas where those wells were constructed. Suggestions for the modern drilling data base systems in Russia were based first on the KTW drilling data.

References

[1] Gelfgat, Ya. A. and S. P. Maksimov, "Stratigraphic Well Drilling." *Mining Encyclopedia, Volume 3*. Moscow: Soviet Encyclopedia Publishing, 575, 1987.

[2] Gelfgat, Ya. A., Yu. S. Vasilyev, A. V. Orlov, F. N. Fomenko, et al., *Technique for KTW Drilling*, Proceedings of VNIIBT, Issue XXXXIII. Moscow: VNIIBT Publishing, 1968.

[3] Gelfgat, Ya. A., Yu. S. Vasilyev, B. A. Vasilyev, A. V. Orlov, F. N. Fomenko, et al., *KTW Drilling Technique*, Proceedings of VNIIBT, Issue 61. Moscow: VNIIBT Publishing, 1971.

[4] Vasilyev, Yu. S., Yu. S. Zmozhin, Yu. D. Semenov, G. V. Podkolzin, V. A. Kildibekov, and Ya. A. Gelfgat, *Technique for KTW Drilling*, includes 15 appendices with techniques, developed by Heads of corresponding VNIIBT departments, Moscow: VNIIBT Publishing, 1976.

[5] Baidyuk, B. V., Zaretsky, et al. *Methodical Guidance for Determining and Utilization of Rock Characteristics for Drilling*. RD 39-30679. Moscow: Minnefteprom publishing, 1983.

[6] Baidyuk, B. V., *Physical-and-Mechanical Fundamentals of Well Drilling Processes*. Moscow: Gazprom Publishing, 1993.

[7] Eigeles, R. M. and R. V. Strekalova, *Well Drilling Processes Calculation and Optimization*. Moscow: Nedra Publishing, 1977.

[8] Dmitriyev, V. N., "Prompt Optimization of Drilling Exploratory and Prospective Wells for Oil and Gas." Synopsis of a PhD Thesis, Moscow Institute of Oil and Gas, 1986.

[9] Fedorov, V. S., *Development of Drilling Practices Design*. Moscow: Gostoptekhizdat Publishing, 1958.

[10] Gulizade, M. P., et al. "Method of Determining Mechanical Drilling Model Coefficients and Drilling Practice Parameters' Limitations for Field Conditions." *Scientific Reports*. Baku: Azneftekhim Publishing, No.1, 1976.

[11] Brevdo, G. D. and K. Gersh. "Drilling Practice Parameters Optimization." *Survey of Information Bureniye (Drilling) Magazine* VNIIOENG Publishing, Moscow, 1968.

[12] Galle, E. M. and H. B. Woods, "How to Calculate Bit Weight and Rotary Speed for Lowest Cost Drilling," *Oil and Gas Journal*, Volume 58, issue 46, 1960.

[13] "Practical Ways to Find Proper Bit Weight and Rotary Speed," *Oil and Gas Journal*, volume 58, issue 47, 1960.

[14] "Best Constant Bit Weight and Rotary Speed." *Oil and Gas Journal*, volume 61, issue 41, 1963.

[15] Young, F. S. "Computerized Drilling Control." *Journal of Petroleum Technology*. volume 21, issue 4, 1969.

[16] Bourgoune, A. T. and F. S. Young, "A Multiple Regression Approach to Optimal Drilling and Abnormal Pressure Detection," *SPE Journal*, August 1974.

[17] Bingham, M. G., "A New Approach To Interpreting Rock Drillability," *Oil and Gas Journal*, volume 62, issues 44–52, volume 63, issues 1–3, 1964–1965.

[18] Fingerit, M. A. *Proper Procedure of Cone Bit Running*. Moscow: Nedra Publishing, 1965.

[19] Volgemut, E. A., M. G. Eskin, V. Kh. Isachenko, et al. *Bit Feed Mechanism for Drilling Oil and Gas Wells*. Moscow: Nedra Publishing, 1969.

[20] Kozlovsky, Ye. A. and V. M. Pitersky, *Results of Scientific and Research Work for Drilling Operations Optimization*. Moscow: VIEMS Publishing, 1982.

[21] Gulizade, M. P., et al. *Modified Simplex Method for Solution of Drilling Optimization Problems*. Collection of Proceedings. IX Series No. 2. Baku: Azineftekhim Publishing, 1979.

[22] Rodgers, C. D. and I. A. Fowler, "Drilling Optimization Searching and Control Method." Pat. 4.195.699 (USA) MKU E21B3/06, 1980.

[23] Edelberg V. "Drilling—Cost Charts Mean Cheaper Wells," *Oil and Gas Journal*, Volume 59, issue 16, 1961.

[24] Hill, Tom H. and Gary Lee Jr. "Drilling optimization: If it ain't broke, fix it!" *World Oil* 220, March 1999.

[25] Dyukov, L. M., V. I. Volkov, and Yu. D. Semenov, *Rock Drillability Evaluation Using Data from Geophysical Survey Revisited.* Proceedings of VNIIBT, issue XIV Moscow: Nedra Publishing, 1965.

[26] Gelfgat, Ya. A. and M. A. Alexandrov, "Overall Drilling Rate and Cost per Foot as Criterion of Drilling Process Economic Efficiency." *"Neftyanoye Khozyaistvo" (Oil Economy and Management) Magazine* 1. Moscow: Nedra Publishing, 1972.

[27] Gelfgat, Ya. A., A. V. Orlov, G. V. Finkelshtein, and V. V. Cherkayev, *Establishing Certain Empirical Dependencies of Bit Performance Results from Drilling Parameters in the Field Conditions Revisited,* Proceedings of VNIIBT, issue IX. Moscow: Gostoptekhizdat Publishing, 1963.

[28] Dyukov, Volkov, and Semenov, 1965.

[29] Gelfgat, Orlov, Finkelshtein, and Cherkayev, 1963.

[30] Orlov, A. V, *Determining Optimum Combination of Bit Weight and Rotational Speed for Deep Well Drilling,* Proceedings of VNIIBT, issue XIII. Moscow: Nedra Publishing, 1964.

[31] Gelfgat, Orlov, Finkelshtein, and Cherkayev, 1963.

[32] Orlov, 1964.

[33] Gelfgat, Vasilyev, Orlov, Fomenko, et al., 1971.

[34] Vasilyev, Zmozhin, Semenov, Podkolzin, Kildibekov, and Gelfgat, 1976.

[35] Gelfgat, Vasilyev, Vasilyev, Orlov, Fomenko, et al., 1971.

[36] Gelfgat, Ya. A., "The Effect of Round Trips Speed on Optimum Drilling Parameters." *Neftyanoye Khozyaistvo (Oil Economy and Management) Magazine* 12, 1966.

[37] Barshai, G. S., Ya. A. Gelfgat, A. Z. Romanov, *Turbodrilling without Pulling Out Drill Pipes*. Moscow: Nedra Publishing, 1967.

[38] Orlov, 1964.

[39] Barshai, Gelfgat, Romanov, 1967.

[40] Finkelshtein, G. M., *On Effect of Drilling Parameters on Power Characteristics and Overall Performance of Blade Bits in High Plasticity Low Abrasive Rock*. Proceedings of VNIIBT, issue XXVIII Moscow: VNIIBT Publishing, 1971.

[41] Finkelshtein, G. M., "Planning of Experiment for Determining Basic Dependencies Coefficient for Drilling Model." *Technique for KTW Drilling*. Attachment 10, Section B. Moscow: VNIIBT Publishing, 1976.

[42] Palii, P. A. *Drilling Engineer's Handbook*. Volume 1, Chapter 5. Moscow: Nedra Publishing, 1973.

[43] Finkelshtein, 1971.

[44] Ibid.

[45] Finkelshtein, 1976.

[46] Ibid.

[47] Gelfgat, Ya. A., A. V. Orlov, G. M. Finkelshtein, G. S. Sharutin, and N. N. Yadullayev, *Summarized Results of Experimental-Demonstration Well Drilling in the Karadag-Damba Field*. Proceedings of VNIIBT, issue XIV Moscow: Nedra Publishing, 1965.

[48] Ibid.

[49] Ibid.

[50] Gelfgat, Ya. A., L. M. Dyukov, V. I. Volkov, Yu. D. Semyonov, and N. P. Levchenko, *Results of Test Well Drilling in the Samara Region*, Proceedings of VNIIBT issue XVII. Moscow: Nedra Publishing, 1967.

[51] Barshai, G. S. and N. I. Buyanovsky, *Theory and Practice of Turbodrilling*, Moscow: Gostoptekhizdat Publishing House, 1961.

[52] Rebinder, P A., K. F. Zhigach and A. A. Shreiner, *Hardness Reducers for Drilling Operations*. U.S.S.R. Academy of Science Publishing, 1944.

[53] Gelfgat, Dyukov, Volkov, Semyonov, and N. P. Levchenko, 1967.

[54] Gelfgat, Ya. A., F. M. Fomenko, V. I. Kurepin, B. I. Abyzbayev, V. F. Sabayev, *Study of Electrodrilling Practices for Drilling through Crystalline Basement in Well No. 2000 in the Bashkiria Region*, Proceedings of VNIIBT issue XIX , Moscow: Nedra Publishing, 1968.

[55] Fomenko, F. N. *Electrodrills for Drilling Oil and Gas Wells*. Moscow: Gostoptekhizdat Publishing, 1961.

[56] Volfson, V. I., Ya. A. Gelfgat, A. V. Orlov and Ye. G. Chervonsky, *Results of Wells Drilling Using 161 mm Bits*. Proceedings of VNIIBT issue XIV Moscow: Nedra Publishing, 1965.

[57] Gelfgat, Ya. A., *On New Operational Procedures for Drilling Deep Wells and Productive Horizons*. Proceedings of VNIIBT, issue XLI. Moscow: VNIIBT Publishing, 1978.

[58] Vasilyev, Yu. S., Yu. S. Zmozhin, E. P. Kaidanov and V. A. Kildibekov, *On Possible Reason for Inadequacy of Bit Performance Results Representation Using Standard Empirical Dependencies*. Proceedings of VNIIBT, issue XLI. Moscow: VNIIBT Publishing, 1978.

[59] Vasilyev, Yu. S., V. P. Kaidanov, and Yu. Yu. Nikitin, *Analysis of Drillstrings with Dividers Dynamics*, Proceedings of VNIIBT, issue XLI, Moscow: VNIIBT Publishing, 1978.

[60] Vasilyev, Yu. S., Ya. A. Gelfgat, E. P. Kaidanov, and Yu. Yu. Nikitin, *Test Studies of Effect of Dynamic Load Non-Linear Pattern on Drilling Results*. Proceedings of VNIIBT, issue XLI. Moscow: VNIIBT Publishing, 1978.

61 Vasilyev, Kaidanov, and Nikitin, 1978.

62 Perov, A. V., "Fruitful Collaboration." *Bureniye Magazine* 6. Moscow: VNIIOENG Publishing, 1978.

63 Gelfgat, Ya. A., G. V. Rogotskii, V. B. Razumov, S. V. Solomennikov and A. L. Kunakh, *Application of Seismic Method for DHM Direction Finding During Directional Wells Turbodrilling in Fields Operated by Orenburgneft*, Proceedings of VNIIBT, issue XLI Moscow: VNIIBT Publishing, 1978.

64 Gelfgat, Ya. A., F. N. Fomenko, and A. U. Yafarov, *Mathematical Statistics Methods for Determining Optimum Drilling Parameters for Fields in Turkmenistan*, Proceedings of VNIIBT, issue XXVIII. Moscow: VNIIBT Publishing, 1971.

65 Gelfgat, Ya. A., "Selection of Efficient Applications for Various Drilling Methods Revisited." *Oil Industry Magazine* 12, Moscow: Nedra Publishing, 1981.

66 Gelfgat, Ya. A., A. V. Orlov, and F. N. Fomenko, "On Selection of Efficient Applications for Various Drilling Methods." *Neftyanoye Khozyaistvo Magazine* 11. Moscow: Nedra Publishing, 1974.

67 Millheim, Keith, "Proactively Managed Drilling Operations Optimize Company Performance," *Oil & Gas Journal*, September 17, 2001.

CONCLUSION

We hope that Volume 1 of this book provided thorough discussion of drilling technologies in the former U.S.S. R. and Russia. These three chapters covered the basics in historical, economical, and technical aspects of drilling technologies development in the FSU. The availability of downhole motors technology along with drilling optimization techniques provided the foundation for advances in drilling processes development. These include cluster directional drilling, horizontal and multilateral wells construction, drilling of super-deep wells, using air, form and aerated mud techniques, and retractable drillbit technology. These technologies are covered in detail in Volume 2, Chapters 4, 5, 6, and 7.

ACRONYMS

AC	alternating current
ADP	aluminum drillpipe
AHP	abnormal high pressure
ASME	American Society of Mechanical Engineers
BHA	bottomhole assembly
CCS	complete coring system
CDR	commercial rate of drilling, meter per rig-month (overall drilling rate)
CIS	Commonwealth Independent States
CT(D)	coiled tubing (drilling) system
DC	direct current (electric motor)
DHM	downhole motor
DOE	U.S. Department of Energy
DP	Drillpipe
DPS	dynamic positioning system
DRC	drillcollar
DS	Drillstring
D/S	drill ship
DT	deflecting tool
DWPP	drill without pulling out the pipe
EDM	electric downhole motor (electrodrill)
EKTB	Experimental Turbine Drilling Bureau
EM	electromagnetic (signal)
EOR	enhanced oil recovery
ESP	electric submersible pump
ERD	extended reach drilling
ERW	extended reach well
EUR	expandable underreamer

FPH foot per hour

FSU former Soviet Union countries

GKNT State Committee on Sciences and Technology of the U.S.S.R.

GOSPLAN State Planning Committee of the U.S.S.R.

GOST State standard both in U.S.S.R. and Russia

HBS hydro-braking system (or stages) in a turbodrill

HDHM hydraulic downhole motor

HH hydraulic hammer

HTHP high temperature and high pressure

H_2S hydrogen sulfide

IADC International Association of Drilling Contractors

ID inside diameter

ISM drag type drill bit with diamond composite inserts (design of the Institute of Super-hard Materials, Kiev, Ukraine

JSC joint-stock company

KTW key test (technological) well

LBF pounds-force

MEI Maurer Engineering Incorporated

MIE military-industrial establishment

MINKh (or MING) Oil and Gas University (later Academy) in Moscow

MTBF mean time between failures

MWD measurement while drilling

NIPI Regional Scientific Research and Production Institute in FSU

NPU oil production unit (department) within the oil company in FSU

OD outside diameter

ODP Ocean Drilling Program

PDC polycrystalline diamond compacts

PDM positive displacement motor

POOH pulled out of hole

R & D Research and development

RB retractable drillbit

RBHA retrievable BHA

RDCB rotary drilling core barrel

RKh "fishtail" drag type drill bit

RPM (rpm) revolutions per minute rotational speed

ROP rate of penetration

RPS rapid piston sampler

RTB rotary-turbine drill or rotary-turbodrill unit

RR retractable reamer

RRA, RRB and RRV different types of EURs

SAMT steering and angle measuring tool

SG (or SD) super-deep borehole (used for well numbers)

SKB Special Design Bureau

SKTBE Special Design and Technological Bureau of Submersible Electromotors including Electrodrills

SPE Society of Petroleum Engineers

STE cable telemetric system developed for electrodrilling

STT cable telemetric system developed for turbodrilling

TCI tungsten carbide insert

TD total depth of the well

TMD total measured depth

TMDB downhole motor coring

TVD total vertical depth

TRB tri-cone retractable drillbit

UVU casing cutting universal tool

VNIIBT All-Union Scientific and Research Institute of Drilling Technology

WOB weight on bit

WOC waiting on cement

WWII World War II

CONTENTS OF VOLUME 2

INDEX

A

B

C

D

INDEX

E

F

G

H

I–J

K

L

M

N

O

P–Q

R

S

T

U

V

W–Z

NOTES

NOTES

NOTES

NOTES

NOTES

NOTES

NOTES

NOTES

NOTES

NOTES

NOTES

NOTES

NOTES

NOTES